# 3D Printing Book

*"Mastering 3D Printing Techniques: Advanced Skills for Intermediate Creators"*

Zack Ross

# Table of Contents

# Introduction

## Transitioning from beginner to master

The road from novice to expert in any field is both intricate and captivating. It's a voyage punctuated by growth challenges, revelations, and the unyielding ambition to expand one's horizons. This transformation from a fledgling to an adept isn't just about amassing experience; it's about the cultivation of wisdom, the polishing of abilities, and a profound comprehension of the intricacies of the subject.

1. Setting Out: Novice Horizons

All experts began their journey as novices. This phase is marked by eagerness, inquisitiveness, and perhaps a sprinkle of innocence. Newcomers often tackle issues with an unjaded lens, untouched by the constraints of 'traditional methods'. However, this period also witnessed frequent mistakes, misinterpretations, and a heavy dependence on external advice.

2. Quest for Proficiency

Evolving from novice to intermediary entails a deep dive into the subject. This could manifest as formal training, autonomous research, or experiential learning. Here, the individual starts to unravel the rationale behind the techniques. They begin to discern trends, predict obstacles, and carve out their unique approaches. This stage acts as a connector between mere procedural adherence and a genuine grasp of the craft.

3. Polishing the Craft: Advancing the Journey

The progression from intermediary to seasoned professional is less about novel learning and more about refinement. It revolves around sharpening one's expertise, perfecting methodologies, and curbing inaccuracies. An adept individual can execute, but also conceptualize, modify, and innovate when faced with uncharted scenarios. They start to resonate with the nuances and often produce solutions that seamlessly merge efficacy with sophistication.

4. Mastery: An Ever-evolving Process

Intriguingly, authentic expertise isn't a final destination but an ongoing expedition. Those who achieve mastery understand that knowledge is infinite. Every field is in flux, continually offering new methodologies, innovations, and hurdles. Thus, being a master means staying informed, embracing fresh paradigms, and retaining the modesty to realize that perfection is a moving target.

5. The Gift of Mentorship

A true testament to mastery is the capacity and eagerness to mentor the next generation. Sharing insights not only cements one's grasp but also opens avenues to perceive the craft from diverse angles. Through mentoring, masters often discover more layers to their domain, prompted by new questions, alternative thought processes, and the challenge of simplifying intricate notions for greenhorns.

## 6. Evolving in Mindset

Beyond technical prowess, the evolution from beginner to master demands a significant mental transition. It's a journey toward resilience, persistence, and critical self-reflection. It encompasses welcoming critiques, recognizing areas of growth, and ceaselessly elevating one's own standards. It's the evolution from seeking affirmation to chasing brilliance.

## 7. Nurturing through Community

No evolution from greenhorn to guru occurs in a vacuum. Immersing oneself in a community, be it a recognized group, an informal gathering, or digital platforms, is pivotal. Such communities offer avenues for collaboration, knowledge exchange, counsel, and keeping a pulse on the ever-evolving landscape of the domain.

## In Summation

The metamorphosis from beginner to master is a multifaceted expedition that transcends skill acquisition. It's a journey of transformation, marked by personal evolution, perpetual education, and an enriched perspective of the craft. While the path may be strewn with challenges, the intangible rewards of self-contentment, acclaim, and the chance to make a lasting imprint on the domain are priceless.

## The world beyond basic 3D printing

The realm of 3D printing, once considered an esoteric technology reserved for specific labs, has burgeoned into a familiar entity in contemporary society. With applications ranging from designing intricate adornments to erecting complete structures, the momentum behind this technological marvel continues to accelerate. As it becomes more mainstream, it's essential to explore the advancements stretching beyond its foundational uses. We're on the cusp of a transformative period where 3D printing will redefine how we approach manufacturing, artistry, and engineering.

Diversifying Printing Materials

One monumental stride in 3D printing is the broader palette of materials at our disposal. Moving past the conventional plastics, such as PLA or ABS, there's a burgeoning interest in harnessing advanced polymers, metals, ceramics, and even biologically derived materials.

Printing with metals, for example, is rapidly becoming a staple in sectors like aerospace and automotive industries. Utilizing metals like titanium alloys provides the advantage of producing components that are both lightweight and robust. Moreover, the emergence of composite materials, combining the traits of multiple materials, allows manufacturers to customize products to achieve desired thermal, mechanical, or electrical properties.

## Bioprinting: Pioneering Medical Approaches

Among the most captivating advances is the domain of 3D bioprinting. Employing bio-inks crafted from living cells, there's potential to fabricate tissues, full organs, and other biological entities. While still in its nascent stages, the ramifications are monumental. Envision a scenario where the need for organ donors becomes obsolete, offering patients bespoke organs, perfectly aligned with their biological specifics, minimizing rejection risks.

## Harnessing Multiple Materials and Hues

Early 3D printing predominantly focused on monomaterial constructs. Contemporary printers, however, are equipped to manage a mix of materials and shades within one printing session. This versatility enhances both the visual appeal and functional capabilities of the printed entity. For instance, crafting a prototype can now seamlessly integrate both stiff and pliable parts, closely mirroring the end product's operational nuances.

## 4D Printing: Beyond Static Constructs

While 3D printing produces fixed objects, 4D printing brings in a dynamic component: time. Leveraging responsive materials, these printed items can morph in shape or function post-production, reacting to external triggers like temperature or light. This innovation hints at future prospects like structures that self-assemble or devices that auto-adjust based on ambient conditions.

## Precision Printing: Micro and Nano Domains

Achieving exactness at microscopic magnitudes is now feasible thanks to micro and nano-scale 3D printing. This meticulousness is paramount in fields like electronics, emphasizing miniaturized components, or in biomedical research, where tiny structures can guide cellular activity.

## Environmental Responsibility in 3D Printing

With 3D printing's proliferation, the imperative for eco-consciousness becomes evident. Cutting-edge initiatives in this sphere are promoting the use of repurposed materials. Efforts to transform consumer waste into printable resources are underway. Simultaneously, there's a growing emphasis on biodegradable resources, promoting sustainable and environmentally friendly prints.

## Software Progressions

Not just limited to hardware upgrades, the software underpinning 3D printing is also evolving. Modern design platforms now harness the power of machine learning to refine designs, assuring material efficacy and structural soundness. Predictive software tools allow creators to anticipate how printed elements will react under varying scenarios, sidestepping the outdated guesswork methodology.

## In Closing

Peering beyond the elementary facets of 3D printing reveals a dynamic landscape, signaling not just technological growth but a significant shift in perspective. From the raw materials to our conceptual framework around production and design, 3D printing serves as an emblem of human ingenuity. As we venture further, it's evident that the only boundary to this technology's potential is the scope of our collective creativity.

## Setting the context: what this book offers

Navigating the world of technology can sometimes feel like traversing an endless maze. Detailed nuances can get lost in dense academic papers, while the basics might not be evident in expert treatises. This work is designed to be the bridge—providing both an in-depth analysis and a friendly guide for newcomers. By establishing the groundwork upfront, we prepare our readers for the rich journey that unfolds in the chapters that follow.

This book's primary aim? Equip readers with a comprehensive understanding. Those new to the topic will find the initial chapters invaluable, laying the groundwork and easing them into more complex subjects. Those familiar with the subject are also in for a treat. Sometimes, revisiting the basics can shine a light on overlooked facets or deepen one's comprehension.

The Journey from Basics to Advanced Topics

While the book does establish foundational knowledge, it doesn't stop there. We journey into the complexities, dissecting recent breakthroughs, and emerging techniques, and speculating on future paths. The scope of our discussion is vast, ensuring a holistic view of the topic.

Hands-on Learning Approach

Recognizing that true knowledge comes from doing, the book is infused with interactive elements. Readers will find exercises, practical assignments, and scenarios sprinkled throughout, encouraging them to apply what they've learned and solidify their understanding.

Gleaning Insights from Industry Leaders

Adding layers to our discussion, insights from industry forerunners have been integrated. Through candid interviews and specialized sections, readers get a peek into the minds driving the industry forward, offering a well-balanced perspective.

Decoding the Technical Lingo

Every domain has its own language. However, technical terminologies shouldn't be barriers. With this in mind, we've made every effort to clarify industry-specific jargon. Alongside a thorough glossary, any new term introduced is promptly clarified in layman’s terms.

Peering into the Future

While a good portion of this work discusses the present landscape, we don't stop there. An essential aspect of our narrative is to gaze ahead, pondering over emerging patterns, potential developments, and their broader societal repercussions.

## Wrapping It Up

Think of this book as a guided tour. It's designed to lead you from the comforting realms of the known into the exciting territories of the unknown. By setting the context from the get-go, we're ensuring that you, our dear reader, are primed to extract the most from this enriching journey. Dive in, for a world of knowledge awaits!

# Chapter One

## Delving Deep: An Advanced Overview of 3D Printing Technologies

### A refresher on primary technologies and their advancements

Navigating the modern technological era, one can't help but marvel at the rapid progress shaping our world. As we journey deeper into this age of innovation, there are foundational technologies that stand out, continually redefining our future. This article endeavors to provide an updated overview of these transformative technologies, highlighting their recent breakthroughs and evolution.

1. Artificial Intelligence (AI)

Born from the desire to replicate human cognition in machines, AI has surged ahead in recent years. Its remarkable progress can be attributed to powerful computing capabilities coupled with the availability of big data. Sub-disciplines like machine learning and deep learning have empowered systems to decipher complex patterns, optimize decisions, and even anticipate upcoming trends. Its influence spans from precision medicine to stock market predictions.

2. Internet of Things (IoT)

The notion of interconnected devices, IoT, is about embedding intelligence into everyday items, allowing them to communicate seamlessly. Though not a novel concept, its practical applications have grown exponentially. Advancements in sensor technology, more affordable components, and the emergence of superior connectivity like 5G have propelled IoT's influence. Today's smart homes, interconnected transport systems, and even intelligent urban environments owe their existence to IoT.

3. Blockchain

Initially rooted in cryptocurrency mechanisms, blockchain represents a decentralized approach to record-keeping. Every activity becomes a 'block' that's securely linked in a chain of previous and subsequent actions. Its brilliance lies in its unalterable transparency. Beyond digital currency, blockchain's versatile applications now encompass areas like logistics, medical record systems, and even transparent electoral procedures.

4. Quantum Computing

While conventional computing hinges on bits, quantum computing introduces the world to quantum bits or qubits, capable of embodying multiple states concurrently due to superposition. This multiplicity enables them to conduct numerous computations in tandem, showcasing immense potential. Though we're only scratching the surface, evolving quantum algorithms and equipment are narrowing the gap towards feasible quantum computing solutions.

5. Augmented Reality (AR) and Virtual Reality (VR)

AR and VR are reimagining our interaction with digital information. VR immerses its users in a simulated realm, whereas AR enhances our real world with digital augmentations. Originally popularized by gaming enthusiasts, these technologies' horizons are expanding. Enhanced hardware and increasingly immersive software simulations are paving the way for applications in sectors like education, healthcare, and property.

6. 5G Technology

Representing the upcoming phase in mobile connectivity, 5G boasts speeds that potentially outpace 4G by a hundredfold. But its essence transcends speed. Its true merit lies in real-time device and system connectivity, which is instrumental for innovations like self-driven cars and remote medical consultations. With minimal lag and a capacity for a plethora of simultaneous connections, 5G is poised to underpin countless technological strides in the forthcoming years.

7. Biotechnology

Encompassing a spectrum of technologies, biotechnology merges biological insights with technological solutions. Recent breakthroughs, be it the precision of CRISPR for gene modification or bioinformatics' role in pharmaceutical innovations, underscore its transformative potential. Merging biology and tech, opportunities arise in diverse areas, from individualized treatment plans to sustainable energy sources.

8. Nanotechnology

Delving into the microscopic domain, where one nanometer represents a billionth of a meter, nanotechnology promises groundbreaking solutions across industries. Whether it's health, where minute particles can ensure targeted medicine delivery, or electronics, with transistors at the nanoscale challenging traditional limits, nanotechnology's promise is steadily coming to fruition.

In Summary

The leaps in technological growth witnessed in recent times overshadow many past milestones. As these cornerstone technologies further mature, their intersections and collaborative potential will pave the way for more revolutions. This article, then, is both a nod to human innovation and an exciting preview of the limitless prospects ahead.

## Hybrid techniques and emerging 3D printing methods

The additive manufacturing realm continues to surprise us with its rapid progression. Within this dynamic landscape, the birth of hybrid technologies and unconventional 3D printing methods is ushering in a new era. These methods synergize traditional manufacturing's strengths with 3D printing's adaptability, setting the stage for breakthroughs in diverse sectors.

## Blending the Old with the New: Hybrid Approaches

Hybrid methodologies bridge the world of 3D printing with conventional subtractive methods, like milling, leading to a more comprehensive approach to producing components.

Combined Machining Processes: Some avant-garde machines feature integrated milling apparatuses. These devices, after laying down a print layer, can refine it, enhancing precision. This eliminates additional finishing steps and results in components with superior finishes and intricate details.

Layer-wise Material Deposition: This approach marries injection molding techniques with 3D printing. The result? The capability to deposit varied materials during a singular print job. This is crucial when fabricating parts with composite structures or when different sections of an item need different material attributes.

## New Horizons in 3D Printing

Two-Photon Lithography (2PL): While traditional lithography is a known entity in the 3D printing world, 2PL introduces an innovative twist by employing high-speed lasers. These lasers can solidify material with nanoscale precision, paving the way for developments in sectors like nanomedicine, electronics, and advanced optics.

Smart 3D Printing: By weaving artificial intelligence into the fabric of 3D printing, this method can enhance designs, foresee potential printing issues, and adjust print settings on the fly. This ensures higher-quality prints and minimizes resource wastage.

Pellet-Based Extrusion: Moving away from the conventional filament or powder-based approach, this method utilizes molding-compatible pellets. This enables quicker print times and introduces a wider palette of usable materials, including environmentally friendly recycled variants.

Sound Wave-Assisted Printing: A revolutionary method, it employs acoustic waves to precisely place minuscule particles within a medium, allowing for the creation of complex patterns pivotal for cutting-edge electronics and medical innovations.

4D Printing: An evolution of 3D printing, 4D printing crafts parts that undergo transformations post-production in response to external triggers, such as temperature or water. This reactive behavior, made possible by state-of-the-art materials and design schemes, has potential applications in numerous fields including aerospace and biomedical engineering.

Varied Material Printing: This technique, rather than maintaining a consistent material throughout, varies material composition during printing. This produces components with differing properties like flexibility or heat resistance, tailored to specific functional requirements.

Impacts and The Path Forward

The innovative strides in 3D printing have real-world consequences. For instance, the automotive and aerospace sectors can harness lightweight, durable components, contributing to greener solutions. The medical field can anticipate a revolution in personalized implants and prosthetics.

Yet, these breakthroughs come with their set of challenges. Merging various techniques calls for a reevaluation of equipment design and streamlined workflows. The success of these methods often hinges on developing or sourcing materials with bespoke properties. Furthermore, the workforce needs to be trained to harness these advancements efficiently.

Safety and regulatory aspects also come into play with these technological evolutions. Establishing consistent processes, assuring product quality, and understanding the ramifications of new materials are all critical areas that warrant rigorous scrutiny.

In wrapping up, the fusion of age-old and novel manufacturing practices is shaping a transformative phase in the production world. The potential of these pioneering 3D printing methods can redefine sectors and alter our perceptions of design and fabrication. As we navigate this transformative phase, harnessing these innovations can pave the way for a future ripe with unprecedented opportunities.

## Niche methods specific to materials and industries

3D printing's technological spectrum has broadened considerably, far beyond the initial confines of basic plastic layering. Various materials present unique challenges and necessitate specialized printing approaches. Furthermore, specific industries have particular needs that standard printing methods cannot address. An exploration of these specialized techniques highlights the breadth and adaptability of 3D printing capabilities.

Techniques Tailored for Materials

Selective Laser Melting for Metal: While metals offer durability, they present unique printing challenges. SLM addresses this by using potent lasers to merge metal powders one layer at a time. Such an approach has transformed industries like aerospace, where achieving strength without additional weight is crucial.

Ceramic Crafting: Ceramics undergo a dual-phase process in 3D printing: first shaping, then sintering. Techniques like robocasting dispense a ceramic paste in detailed patterns, which is later subjected to heat for binder removal and particle consolidation. This method proves invaluable in the biomedical sector, especially for crafting complex bone support structures.

Glass Formation in 3D: While not widespread, certain methods, such as stereolithography, can manipulate specially designed glass-infused resins. After the printing phase, these structures undergo a process similar to sintering to gain transparency.

Approaches Suited for Specific Industries

Bioprinting in Healthcare: One of the groundbreaking subfields of 3D printing is bioprinting, focusing on creating layers using live cells. Employing bio-inks, mixtures of cells, and supportive substances, intricate constructs like tissues or even organs may soon be a reality. The vision is to have bioprinters as essential equipment in medical facilities.

Rapid Sintering for Bulk Production: While 3D printing is not always synonymous with speed, techniques that employ infrared for quick material processing have emerged. Such methods cater especially to industries requiring large-scale part manufacturing.

Electron Beam Fusion for Aerospace Components: The aerospace sector's requirements for precision and resilience have led to the development of electron beam fusion. Here, metal powders are fused using electron beams, providing the robustness vital for aerospace parts.

3D Printed Electronics: By merging conductive materials with standard printing substrates, some 3D devices can craft electronic circuits alongside physical constructs. This innovation has significant implications, especially in the electronics sector, where compact device design is paramount.

## To Sum Up

While generic 3D printing techniques hold their value, specific materials and sectors often need more nuanced, dedicated methods. These specialized techniques are the forefront of additive manufacturing advancements, aligning closely with the unique requirements of various domains. As these methods evolve, the potential of 3D printing will further solidify, paving the way for innovations across multiple industries.

# Chapter Two

## Advanced Materials and Their Capabilities

### Exotic and hybrid materials: metals, ceramics, and more

Additive manufacturing's horizon has seen an incredible expansion with the advent of new, sophisticated materials. While plastic filaments once dominated the 3D printing scene, the landscape has evolved to encompass a range of metals, ceramics, and multifaceted composites. This deep dive aims to shed light on these modern materials, emphasizing their distinct attributes, utilization, and the game-changing influence they introduce.

Metallic Frontiers in Additive Manufacturing

Metal materials have carved a niche in the 3D printing world, thanks to their inherent robustness and longevity. An assortment of metals and alloys now allows the fabrication of detailed designs formerly considered intricate.

Titanium: Renowned for its impressive strength-to-weight ratio and biocompatibility, titanium is a preferred choice in sectors like aerospace and health. Processes such as Direct Metal Laser Sintering (DMLS) enable the creation of custom implants and aero-components using titanium.

Stainless Steel: Marrying durability with resistance to corrosion, stainless steel has found applications in sectors like automotive, industrial tooling, and jewelry. Methods like bound metal deposition and binder jetting are frequently used with this material.

Precious Metals: The ability to 3D print with gold, silver, and platinum has revolutionized the jewelry domain. Such metals can be transformed into bespoke, detailed designs, pushing traditional crafting boundaries.

Ceramics: The Fusion of Fragility and Fortitude

Ceramics, characterized by their hardness, heat resilience, and brittleness, offer both challenges and prospects in 3D printing.

Technical Ceramics: Alumina, zirconia, and similar materials pave the way for applications across the automotive, aerospace, and electronic industries. High-resolution prints with these materials are achieved through processes like lithography-based ceramic fabrication.

Bio-Ceramics: Ceramics that promote bone integration have carved a niche in the healthcare sector. Materials such as hydroxyapatite are pivotal in dental and skeletal implants.

Composites and Hybrid Materials: Merging Best Traits

The realm of 3D printing materials has seen the emergence of composites, which integrate attributes from different sources, offering unparalleled versatility.

Carbon Fiber Infused Polymers: Integrating carbon fibers' robustness with polymers' adaptability results in materials ideal for sectors like aerospace and automotive, combining lightweight properties with sturdiness.

Filaments Enriched with Graphene: Merging standard plastics with graphene's extraordinary properties leads to heightened thermal and electrical capabilities, promising innovative applications in diverse fields.

Metal-Polymer Fusions: Marrying metal powders with thermoplastics results in unique materials that bring together metal and plastic benefits. Post-printing processes typically involve sintering the metal to leave a pure metal structure.

Concluding Thoughts

The proliferation of exotic and composite materials has undeniably enriched the 3D printing spectrum. From crafting components that withstand harsh environments to meeting specific medical needs, the potential is vast. The integration of these materials with cutting-edge printing methods has created a dynamic synergy.

Yet, challenges persist. Issues related to material procurement, expense, and the complexities of the printing and post-processing stages often arise. Nevertheless, continued research promises solutions, unlocking these materials' full capabilities.

In essence, the palette of 3D printing materials has grown diverse and complex. With the inclusion of ceramics, metals, and hybrids, the future appears replete with promise, with these materials poised to drive further innovation in the world of additive manufacturing.

# Material science behind advanced 3D printing materials

As the realm of 3D printing evolves, a critical factor driving this transformation is the progression in material science. While the early days of additive manufacturing saw dominance by basic plastics, now we witness a myriad of materials showcasing vast properties. A dive into these innovative materials provides insights into their capabilities and potential.

3D Printing with Metallic Constituents

The utility of metals, characterized by their robustness, conductivity, and flexibility, has expanded within additive manufacturing. Recognizing their atomic structure and transitional phases during printing processes is paramount.

Titanium and Its Combinations: Renowned for its high strength-weight metrics and exceptional biocompatibility, titanium's distinctive structural phases enhance its mechanical attributes. The swift heating and cooling cycles in 3D printing can instigate changes in its micro-architecture, impacting the end material properties.

Stainless Steel Mixtures: With iron and chromium at its core, stainless steel's microstructure can be ferritic or austenitic. Powder-bed fusion is a popular 3D printing technique for this metal, and understanding its melt-solidify cycle is essential.

## Ceramic Materials in Printing

Ceramics, essentially inorganic and non-metallic compounds, bring thermal resilience, rigidity, and electrical isolation to the table in 3D printing.

Zirconia ($ZrO_2$): Zirconia stands out with its resistance to wear and robustness. During printing, stabilizing its tetragonal phase is critical for harnessing its mechanical strengths.

Alumina ($Al_2O_3$): Characterized by its hardness and insulative properties, alumina's crystalline makeup and granular boundaries play roles in its post-printing characteristics.

## Polymeric and Curing Resins

Polymers, spanning both thermoplastics and curing resins, are cornerstones in 3D printing.

Poly(lactic acid) (PLA): Originating from sustainable sources, PLA, a thermoplastic, displays unique traits based on its crystallinity and transition temperatures.

Curing Resins: Employed in techniques like stereolithography (SLA), these resins undergo a set polymerization when exposed to UV light. Aspects like molecular structure, linkage densities, and activator levels influence their mechanical properties.

## Blended Materials in 3D Printing

The union of different materials can culminate in a substance that encapsulates the strengths of each.

Carbon Fiber-Infused Polymers: These materials consist of a base polymer, often epoxy, enhanced with carbon fibers. It's essential to consider aspects like fiber direction, constituent ratio, and bonding to gauge the composite's performance.

Metal-Centric Composites: Here, a foundational metal like aluminum gets enhanced with ceramic fragments or fibers. Elements such as fragment dispersion, magnitude, and bond strength become vital.

Responsive and Adaptive Materials

These innovative materials possess the ability to adjust or react to external conditions, introducing a new dimension to 3D printing.

Memory Alloys: Primarily comprising metals like nickel and titanium, these alloys can assume their pre-defined form when heated. Their action is rooted in the shifts between their atomic phases.

Electronically Conductive Pastes: These consist of metallic nano-particles embedded within a polymer base, which when 3D printed can act as conductive pathways. Factors like particle dimensions, arrangement, and heat processes define their eventual conductive nature.

Concluding Thoughts

Understanding the deep-rooted material science of advanced 3D printing substances reveals the balance of chemical and physical interactions involved. With their diverse structures, molecular interplays, and behavior, each material introduces unique challenges and avenues in additive manufacturing.

As exploratory efforts intensify, the horizons of what can be achieved with 3D printing continue to broaden. A profound grasp of these materials not only uplifts the caliber of printed artifacts but also catalyzes unprecedented applications, ushering in a transformative era in fabrication and creative design.

## Properties, challenges, and benefits of non-traditional materials

Exploring the cutting-edge field of material science reveals the innovation and development of unconventional materials. While their distinct attributes bring numerous advantages, there are inherent challenges in harnessing their full potential. A deeper examination of these avant-garde materials allows us to understand their essence, their strengths, and the obstacles in their application.

Characteristics of Non-Traditional Materials

Self-repairing Materials: Drawing inspiration from nature's regenerative processes, these materials possess the innate capacity to mend damages or fractures autonomously. Their ability to self-repair stems from their specific chemical makeup or from healing agents encapsulated within.

Phase Altering Materials (PCMs): PCMs are notable for their transition between physical states, mainly solid and liquid, within a specific temperature range. This state alteration is associated with the latent heat's absorption or release.

Aerogels: Often dubbed as 'liquid air,' aerogels stand out for their extremely low density and porous nature. Their structure is achieved by substituting a gel's liquid component with gas.

Piezoelectric Substances: Upon experiencing mechanical pressure, these substances produce an electric current, thanks to their asymmetrical crystalline structure.

Obstacles with Non-Traditional Materials

Production and Scale: The specialized properties of many unconventional materials often lead to challenges in mass production. The techniques required can be complex, or the foundational materials might be in limited supply.

Economic Considerations: Significant investment in research, development, and acquisition can drive up costs. This financial aspect may restrict their widespread use.

Longevity and Robustness: Even though they may exhibit extraordinary attributes, some materials might not maintain their qualities under extended or severe conditions.

Information Deficiencies: As these materials are still subjects of ongoing research, there's more to uncover about them. Such gaps can lead to uncertainties in real-world applications.

Merits of Non-Traditional Materials

Enhanced Performance: Many unconventional materials boast superior thermal, electrical, or mechanical traits when juxtaposed with traditional materials. For instance, aerogels offer unparalleled thermal resistance with minimal weight.

Eco-friendliness: Materials such as bio-derived polymers underscore environmental responsibility by reducing fossil fuel dependence and offering biodegradability.

Stimulating Innovations: The singular attributes of these materials can inspire groundbreaking discoveries. For example, piezoelectric substances are instrumental in converting mechanical forces into electric energy.

Safety and Upkeep: Materials that can self-repair, for instance, can extend a product's functional life, decreasing regular maintenance and bolstering safety.

Energy Utilization Efficiency: Thanks to their heat management capabilities, PCMs can significantly enhance energy conservation in structures or tech devices.

Moving Ahead

The journey with unconventional materials, brimming with promise, is not without its challenges. Industry leaders and researchers are continuously working to tackle these hurdles to broaden their applications. Their overarching mission is to capitalize on the unique traits of these materials to achieve breakthroughs previously considered unfeasible.

The importance of these materials in revolutionizing industries and spearheading novel solutions is gradually coming to the fore. From construction and electronics to aviation and healthcare, their transformative impact is undeniable. Recognizing and harnessing the power of these materials, while comprehending their nuances, can shape a future marked by ground-breaking advancements and sustainability.

# Chapter Three

## Precision Designing for Expert Results

### Advanced CAD techniques and tools

The world of Computer-Aided Design (CAD) has witnessed an evolution like no other, moving from basic drafting functions to intricate design methodologies powered by progressive tools. As industries continue to demand more complex designs, CAD tools have stepped up to the plate, offering a plethora of advanced techniques that pave the way for tomorrow's innovations. Let's delve deeper into these groundbreaking techniques and understand their transformative potential.

Dynamic Parametric Design: Instead of using fixed dimensions, this approach uses adaptable parameters. Such parameters, whether they be dimensions or restrictions, permit alterations to specific design aspects, ensuring corresponding elements adjust automatically. This dynamic adaptability streamlines alterations, paving the way for an efficient design evolution.

Design Generation: Shifting away from conventional methods, this approach requires users to feed their design objectives and criteria into CAD systems. The software then produces a myriad of design alternatives, drawing upon set algorithms and criteria. This technique is invaluable when pursuing the most effective solution from a range of options.

Digital Replication: This method creates a virtual, real-time clone of a tangible object or system. By keeping the digital copy in sync with its physical counterpart, professionals can better monitor, refine, and perfect their designs, ensuring a smoother transition from the virtual realm to physical production.

3D Digitization: By utilizing cutting-edge scanners, objects can be transformed into a digital format, resulting in a detailed point cloud. These points can then be transformed into a 3D model within sophisticated CAD systems, an indispensable asset for enhancing existing designs or reverse-engineering.

Building Information Framework: More commonly utilized within architectural and construction spheres, this framework goes beyond traditional CAD designs, encompassing a myriad of details from spatial aspects to intricate building component properties. It aids in more cohesive and informed decision-making.

Predictive Analysis: Contemporary CAD platforms offer in-depth simulation tools, enabling users to foresee how a product might fare under specific circumstances. Be it analyzing potential wear and tear or understanding thermal dynamics, these predictive tools nip potential design flaws in the bud.

CAM Integration: Seamlessly blending CAD with Computer-Aided Manufacturing (CAM) ensures that designs can easily transition to the production stage. This synergy ensures a smoother manufacturing process, reducing the chances of any discrepancies.

Collaboration in the Cloud: With cloud technology's rise, CAD tools have taken a collaborative turn. Projects can be globally shared, altered, and accessed, fostering a collaborative spirit and fast-tracking project completion.

AR & VR Integration: Marrying CAD with Augmented and Virtual Reality technologies gives designers the unique opportunity to step into their designs. This immersive experience provides a comprehensive view, highlighting areas for enhancement or modification.

Personalized Toolsets: Contemporary CAD platforms empower users with the ability to tailor their toolsets. This customization, driven by project requirements or individual preferences, optimizes the design workflow.

Harnessing AI: Incorporating Artificial Intelligence and Machine Learning within CAD tools is ushering in a new era of design. With the capability to sift through vast data, predict upcoming design shifts, and provide deep insights, CAD tools are now smarter and more intuitive than ever.

To Sum Up

The realm of advanced CAD is at the forefront of design excellence. By capitalizing on these state-of-the-art techniques and tools, industry professionals aren't just building models; they're crafting the future, breaking barriers, and reimagining the boundaries of possibility. As the CAD landscape continues its upward trajectory, it promises a future where designs transcend imagination.

## Designing for mechanical strength and functional parts

In the world of engineering and product development, creating components meant for actual use, especially those that must withstand mechanical pressures, is a critical focus. The challenge is blending design efficiency with durability, a task that demands a deep grasp of materials, design principles, and real-world implications. This exploration delves into the core guidelines for crafting components that are as functionally reliable as they are aesthetically pleasing.

1. Choosing the Right Material: A design's durability often starts with the right material choice. Parameters like tensile resilience, flexibility, hardness, and resistance to wear are pivotal. It's about more than just opting for the toughest material; it's about matching the material to the specific stresses and conditions the component will encounter. For example, a metal might be resistant to pressure but could deteriorate in damp environments.

2. Shape and Force Management: The component's shape and design can heavily influence its mechanical durability. Areas of stress concentration, typically present at sharp edges or sudden geometry transitions, can undermine an otherwise solid design. Implementing curves, steering clear of sharp indentations, and gauging how forces spread across the component can mitigate unexpected weak points.

3. Consistency in Wall Thickness: For many components, from molded pieces to 3D printed items, maintaining an even wall thickness can prevent uneven cooling or shrinking. Different thickness levels might induce internal pressures or distortions

that can undermine the component's overall strength. In the realm of 3D printing, ensuring the walls are sufficiently thick to sustain the model during its creation is key.

4. Joining Methods: How components connect or interlock is a critical design facet. Whether through bolts, welding, or snap mechanisms, the endurance and efficiency of these unions are essential. For instance, using bolts might necessitate strengthened regions or integrated nuts for a solid union.

5. Layer Alignment and Creation Direction: In 3D printing contexts, the placement of a component during its creation can impact its mechanical resilience. 3D printing layers inherently present potential weak points, so coordinating these layers with the force direction can bolster strength.

6. Enhancing Through Post-Processing: Once created, several post-creation techniques can enhance mechanical resilience. Methods such as temperature-based stress relief or surface modifications like peening can elevate resistance to wear and tear.

7. Digital Simulations and Finite Element Analysis (FEA): Advanced digital tools offer designers the capability to preview a component's behavior under defined conditions. FEA can spotlight stress areas, anticipate failure points, and guide revisions in design for optimal performance.

8. Prototype and Test: Simulations offer valuable foresight, but real-world testing remains essential. Creating prototypes and testing them under actual conditions can unveil unanticipated weak spots. Adjusting designs based on these practical outcomes ensures the end product meets real-world demands reliably.

9. Safety Margin Implementation: A foundational principle in design is accounting for a safety margin. If a component should support a 50kg force, designing it to bear up to 75kg provides a cushion for unforeseen variables, material variances, or unexpected situations.

10. Keeping Up with Material Innovations: Continual advancements in the materials science sector bring forth new composites, alloys, and polymers. Staying updated on these breakthroughs can introduce new possibilities for creating lightweight yet robust components.

Wrapping Up

Crafting components with mechanical durability is a multi-layered process. It transcends the realms of mere aesthetics, demanding that when realized, designs meet and exceed their functional expectations in real-world scenarios. Through informed material choices, cutting-edge design tools, and the synergy of theory and practice, engineers and designers can produce components that seamlessly combine form and function.

## Aesthetics, tolerances, and detailing in design

Navigating the intersection of engineering and artistry presents an intricate dance of melding functional precision with visual elegance. Three pivotal facets lie at the heart of this dance: aesthetics, tolerances, and detailing. While each facet has its own unique attributes, they seamlessly converge to determine a product's practicality, appeal, and performance in the real world.

Aesthetics: Crafting a Sensory Experience

Every creation communicates through its visual and tactile elements. Aesthetics are the drivers behind this communication, dictating how users interact with and perceive an item. This interaction extends beyond mere visuals; it's also about the touch, sound, and balance of an object, playing a vital role in evoking emotions and defining perceived worth.

Shape and Shade: The fundamental aspects, from the contours to the palette, lay the groundwork for user interaction. Their synergistic integration creates a coherent and delightful user journey.

Surface and Substance: Beyond the primary visuals, the tactile nature and the chosen material define the sensory feedback. Whether it's the roughness of canvas, the cold of steel, or the cushion of foam, each has its own emotional resonance.

Tolerances: Ensuring Flawless Functionality

While aesthetics captivate, tolerances anchor a product's functionality. They define the permissible variances in design that ensure consistent performance.

Interplay of Components: Essential for parts that move or interlock, precise tolerances are key. They ensure that parts fit seamlessly, without undue wear or space, preserving smooth operation. In sectors like aviation or automotive, such precision is non-negotiable.

Material Dynamics: Given that materials might shrink, stretch, or deform under certain conditions, tolerances must anticipate and counteract these shifts, ensuring sustained operational integrity.

Production Techniques: The manufacturing process chosen inevitably impacts the range of tolerances. For example, while 3D printing might offer bespoke precision, traditional craftsmanship might entail more variations.

Detailing: Nuances that Elevate Design

Detailing is the art of refining, and adding layers of depth to a design. These minute touches, subtle yet profound, enhance user engagement.

Human-centric Design: This might manifest as the arch of a laptop stand, the layout of dashboard controls, or the contour of a handheld device. Such details, when thoughtfully incorporated, amplify user comfort and efficiency.

Interactive Elements: The satisfying snap of a switch, the gentle illumination of a status light, or the muted buzz of a notification are all feedback details that enrich user interaction.

Features for Longevity: Details can also boost product endurance. Aspects like corner reinforcements, protective layers, or anti-fade treatments can extend the life of an item.

Synergizing Aesthetics, Tolerances, and Detailing

The real marvel unfolds in the unison of these elements. A product that nails tolerances but lacks visual allure might be efficient but not enticing. On the flip side, a striking product with flawed tolerances could disappoint functionally. Overlooking detailing can result in items that, while serviceable and stylish, lack that touch of luxury.

The journey of transforming an idea into a tangible entity is multifaceted. Aesthetics, tolerances, and detailing are the pillars that, when harmonized, craft designs that captivate and cater. A meticulous approach to each of these dimensions empowers designers and engineers to produce standout offerings in a dense and demanding market.

# Chapter Four

## Multi-material Printing and Color Techniques

### Setting up and calibrating for multi-material prints

3D printing's dynamic landscape has seen impressive advancements, with multi-material printing standing out as a key innovation. This capability promises the creation of intricate parts that merge various materials, allowing for unique attributes that couldn't be achieved through singular material processes. But to unlock its full potential, attention to detail in setup and calibration is imperative.

Key Considerations in Initial Setup

Printer Suitability: Ensure your 3D printer can accommodate multi-material printing, either through multiple extruders or specific features tailored for alternating materials.

Selecting the Right Materials: It's crucial to pick materials that align in terms of melting behaviors, adhesion characteristics, and temperature responsiveness. Inconsistent material pairings might result in weak bonds or internal tension.

Software Requirements: Utilizing sophisticated software is a must, capable of orchestrating separate pathways for every material and managing their interchange seamlessly.

Essential Calibration Procedures for Multi-Material Outputs

Aligning Extruders: For multi-extruder printers, calibrating each extruder's deposition rate is essential. It's about ensuring consistent output and eliminating potential layer misalignment.

Recommendation: To verify consistency, undertake single-material prints with every extruder individually.

Material-Specific Settings: Dedicate profiles for each material, fine-tuning temperatures, speeds, and other relevant parameters. Misconfigured settings might result in material degradation or bonding issues.

Optimizing Nozzle Positioning: With multiple extruders, it's paramount that all nozzles coexist on the same plane. Minor mispositions can result in one nozzle impeding the print or uneven layer application.

Refining Retraction Parameters: Customize retraction specifics for each material, minimizing material leakage when switching between materials.

Transition Mechanisms: Implement mechanisms to cleanse the nozzle during material transitions, ensuring that each material remains uncontaminated. Techniques include using a wipe wall or a purge tower for residual material disposal.

Managing Transition Areas: There might be blending zones during material switches. Adjusting these zones promises cleaner transitions and minimizes unintended mixed areas in the final piece.

Bonding Trials: Undertake tests to ensure your chosen materials adhere effectively to each other. Proper adhesion is pivotal for the structural integrity of the output.

Consistent Ambient Conditions: It's vital to maintain a stable temperature within the printing environment. Variances can influence the behavior of materials, especially when combined.

Post-Processing Insights: Factor in potential post-production steps, such as stress relief methods, when dealing with multi-material prints.

Potential Hurdles and Solutions

Residue and Material Strings: Transitioning materials can lead to material drippings. This can be managed by optimizing retraction configurations and adopting a consistent purging system.

Layer Separation: If you observe weak bonds between layers, adjust the printing temperatures, check material compatibility, or explore adhesion enhancers.

Variable Material Output: If there's an inconsistency in material flow, consider recalibrating the extruder rates or checking for nozzle obstructions.

Heat Management: Materials with significantly varied temperature needs can pose challenges. Address this by opting for materials with similar thermal profiles or leveraging equipment that can quickly modify temperatures.

Wrapping Up

Embarking on multi-material 3D printing presents an array of possibilities but demands precision and knowledge. With thorough setup and calibration, the process transforms into a powerful tool, ushering in a new era of design and manufacturing innovation.

## Strategies for color consistency and vibrancy

In today's visual-centric era, the emphasis on preserving color vibrancy and consistency cannot be overstated. Whether we're discussing manufacturing, digital artwork, traditional painting, or even 3D printing, striking the right color balance is a harmonious blend of artistry and technique. Below, we unpack some pivotal strategies for maintaining color integrity and luminosity across various platforms.

1. Grasping the Dynamics of Color

More than the light waves that objects emit or reflect, color encompasses the way our vision captures these waves and the subsequent cognitive interpretation. A foundational understanding of color relationships and the color wheel is pivotal for any color-related endeavor.

2. Prioritizing Pristine Raw Materials

Lustrous colors stem from top-tier source materials. Be it inks, digital color shades, or dyes, always prioritize the highest quality. Premium materials typically house richer pigments and minimize unwanted contaminants, leading to more authentic color renditions.

3. Aligning Equipment Settings

For digital practitioners, from graphic artists to 3D printers, the synchronization of equipment is indispensable. Regularly tuning devices like computer screens, printers, and scanners ensure that they mirror colors uniformly. Calibration instruments can offer invaluable assistance in this area.

4. Ambient Influence

The environmental light in which colors are showcased can drastically alter their visual appeal. For example, a shade observed under LED lights might deviate from its appearance in daylight. To maintain color accuracy, particularly in artistic or design spaces, standardized lighting is essential. Lighting that replicates the full spectrum of natural light is usually preferable.

5. Upholding Production Standards

For sectors such as print or fabric, an unwavering production protocol is integral to uniform color results. This encompasses regular equipment upkeep, stable machine parameters, and consistent treatment of materials.

6. Digital Color Systems

In the digital realm, having a grasp on color frameworks, such as RGB for screen display and CMYK for print, is vital. Navigating between these models without color distortion is a skill in itself. Employing universal color schemes, like Pantone, can bridge the gap in color continuity between various projects.

7. Pilot Runs

Before diving into a large-scale color project, always endorse preliminary samples or prototypes. This preparatory step enables any necessary tweaks, ensuring the end product is both vibrant and consistent.

8. Continued Learning

Despite sophisticated tools and streamlined processes, the human touch remains irreplaceable. Periodic training or honing individual skills, encompassing color principles and equipment handling, is the cornerstone of sustained color excellence.

9. Guarding the Color Spectrum

Once a vibrant hue is achieved, its preservation is the next challenge. For tangible outputs, protective sealants or layers can be applied to prevent fading. On the digital front, utilizing durable file types and storage practices guarantees prolonged color quality.

10. Constructive Critique and Refinement

A universal strategy, regardless of medium, is to invite and embrace feedback. This can arise from team members, specialized tools assessing color precision or clients. With each feedback loop, adapt and refine your methodology to enhance the color experience continually.

To Conclude

The essence of color, its ability to resonate emotionally, relay messages, and augment aesthetic appeal, holds unparalleled significance in numerous industries. As tools and tactics evolve, the quest for color perfection remains rooted in blending innovative technology, foundational knowledge, and an intrinsic passion for the world of color. By weaving together the strategies highlighted above, both experts and novices can achieve outstanding color vibrancy and consistency in their respective projects.

## Potential pitfalls and solutions

Venturing into today's complex tech-centric and business-focused landscape can often be likened to navigating a maze. Opportunities and challenges await at every turn. While careful planning can minimize many of these hurdles, stumbling blocks are inevitable. Here, we delve into some frequent complications and the corresponding strategies to tackle them.

1. Gap Between Expectation and Outcome

Challenge: A common hurdle arises when there's a disconnect between what participants expect and the final results. This might be due to vague communication, lack of resources, or ambitious goals.

Remedy: An early phase of in-depth research and candid discussions can prevent such mismatches. Setting regular check-ins or benchmarks can also help to catch and address deviations promptly.

2. Technology Hiccups

Challenge: Systems, regardless of their modernity or dependability, are prone to hitches. These can span from minor software glitches to complete equipment breakdowns.

Remedy: Maintain a proactive IT support system. Periodic system checks, consistent hardware care, and scheduled data backups can circumvent major issues. Furthermore, backup systems and emergency plans can ensure smooth sailing during tech crises.

3. Unplanned Scope Expansion

Challenge: Often seen in projects, there's an unintentional expansion or shift in the scope. This can strain resources, derail timelines, and stress budgets.

Remedy: Set clear parameters for the project from the start. Any shifts or additions should be critically assessed, and involved parties should understand the implications of schedules and costs.

4. Skill and Knowledge Gaps

Challenge: The swift evolution of industries poses the risk of teams lagging in necessary skills or updated knowledge.

Remedy: Promote an ethos of continuous upskilling. Set aside resources for training sessions and workshops. Stay attuned to industry shifts and integrate relevant changes into daily practices.

5. Financial Oversights

Challenge: Mishandling funds can result in overspending, lost prospects, or financial strains.

Remedy: Embrace detailed financial planning and monitoring. Modern financial tools can offer timely insights. Revisit financial strategies considering both immediate needs and future aspirations.

6. Overlooking Stakeholder Interests

Challenge: Disregarding the insights, feedback, or sentiments of crucial stakeholders might lead to reduced enthusiasm, unsatisfactory outcomes, or business tensions.

Remedy: Maintain open communication avenues. Engage stakeholders in pivotal decisions and value their perspectives, reinforcing trust and mutual respect.

7. Legal and Compliance Missteps

Challenge: In an ever-changing regulatory framework, neglecting a compliance update or misjudging a law can bring hefty fines.

Remedy: Collaborate with experts in regulatory compliance. Have mechanisms that periodically check for changes in regulations specific to the industry.

8. Single-Strategy Dependency

Challenge: Overcommitting to one strategy or market heightens vulnerabilities during market fluctuations.

Remedy: Spread across varied strategies and markets. Periodically survey market dynamics and stand ready to adjust or shift strategies accordingly.

9. Not Heeding Feedback

Challenge: Working in silos without external feedback or performance benchmarks may impede progress and innovation.

Remedy: Establish structured channels for feedback, such as customer polls, peer evaluations, or industry research. Treat this feedback as valuable input for growth and refinement.

10. Lack of Crisis Preparedness

Challenge: Unpreparedness during unforeseen challenges can escalate the situation.

Remedy: Draft detailed crisis response plans, detailing potential risks, communication approaches, roles during emergencies, and post-crisis reflections. Regular practice sessions ensure the team's readiness in actual crises.

In Closing

While challenges are part and parcel of any venture, they don't have to culminate in dire outcomes. By employing foresight, methodical planning, flexible approaches, and a dedication to constant growth, these impediments can be transformed into opportunities for advancement, learning, and fortitude. The provided remedies are foundational guides for facing challenges and cultivating an environment of distinction and evolution.

# Chapter Five

# Large-scale and High-resolution 3D Printing

## Adjusting to large format printers

Navigating the world of large format printers introduces a unique set of challenges and opportunities. Specifically tailored for industries spanning from architecture and advertising to manufacturing, these machines provide grand-scale outputs but necessitate an evolved skill set and comprehension, different from conventional printers. Delve into the intricacies of transitioning to these mammoth devices.

Delving Deeper

Superficially, a large format printer might resemble an upscaled desktop printer. However, this perception barely scratches the surface. Their vast dimensions come with more advanced mechanisms, specific software requirements, and detailed calibration processes.

1. Ensuring Optimal Print Quality

When you're dealing with grander scales, even minute imperfections can become glaringly noticeable. It's essential to dive deep into the printer's dpi (dots per inch) settings, ensuring the original design matches it seamlessly for optimal output.

2. Material Diversification

These printers can process a diverse array of materials – from fabrics and films to vinyl and canvas. Recognizing how each substrate reacts with inks and understanding their drying dynamics is crucial to prevent mishaps like smudging or misalignments.

3. Consistent Maintenance

Their expansive nature makes routine check-ups non-negotiable. To avoid issues like printhead clogging and to maximize the machine's lifespan, a consistent cleaning regime is imperative.

4. Mastery Of Software

Operating large format printers typically involves specialized software suites. Continual updates and training can ensure teams remain updated, maximizing design optimization for large outputs.

5. Embracing the Learning Phase

Every novel technology demands adaptation. Some initial challenges, such as color discrepancies or alignment misjudgments, are part and parcel of the journey. Documenting these glitches and deducing their causes will ensure smoother workflows ahead.

6. Prioritizing Safety

With increased capabilities come heightened safety considerations. Ensuring a hazard-free workspace, understanding the printer's power dynamics, and providing ample ventilation can thwart potential mishaps.

7. Financial Aspects

Beyond the initial investment, factors like periodic maintenance, costlier print materials, and potential replacement parts can impact budgets. Proper financial planning, keeping these elements in mind, is crucial.

8. Evolving Design Strategies

A design that's stunning on a smaller scale may not translate as effectively on a grander canvas. Adapting to large format printers means revisiting foundational design elements like spacing, typography, and visual balance.

9. Green Practices

The ecological footprint of these printers, given their energy and material consumption, can be considerable. Incorporating sustainable measures, like energy-efficient operations and using eco-friendly materials, can offset these concerns.

10. Managing Anticipations

While these printers are a gateway to broader horizons, they're not miraculous entities. Setting realistic expectations about output capacities, design potentials, and delivery timelines is fundamental.

Wrapping Up

Transitioning to large format printers is akin to adopting a novel artistic medium. It marries the intricacies of sophisticated equipment with refined design sensibilities. As with any transition, perseverance, ongoing education, and an innovative spirit are keys. With these in hand, the journey into large-format printing can become a rewarding adventure, unlocking a plethora of creative and commercial avenues.

## Achieving microscopic details in high-resolution printing

The realm of high-resolution printing stands at the crossroads of art and modern technological prowess. As the demand for ultra-precise and nuanced products grows across industries, mastering the intricacies and techniques of high-resolution printing is more important than ever.

Delving into High-Resolution Basics

High-resolution printing hinges on a printer's ability to capture and reproduce the most minute of details. This is typically gauged in terms of dots per inch (DPI). A routine office printer might produce outputs in the range of 300 to 600 DPI. In contrast, specialized high-resolution printers can soar to levels of 4800 DPI and beyond, highlighting the finest subtleties.

The Revolution Brought by Inkjet Printers

The heart of this precision-driven output lies in the technology that powers these printers. The inkjet technology, for instance, has been transformative. Using minuscule nozzles, it precisely deposits droplets of ink, each controlled to perfection. Innovations such as the piezoelectric DOD (drop-on-demand) have elevated this accuracy, discharging droplets with impeccable timing.

## Designing with Detail in Mind

The printer's capability is just one piece of the puzzle. The design being printed must be created to exploit this potential. Vector-based graphics, underpinned by mathematical logic rather than pixels, retain their clarity even when scaled up, ensuring razor-sharp results regardless of size.

## The Role of the Substrate

What one prints on is as crucial as the printer used. A pristine, uniformly smooth surface can accentuate the printer's detail-rendering capability. Specialized printing materials, crafted for high-resolution outputs, can further amplify the level of detail.

## The Imperative of Calibration

Given the microscopic scale at which these printers operate, calibration becomes an art form. Periodic maintenance, alignment checks, and printhead upkeep are essential to ensure that every single droplet is precisely placed.

## Navigating Challenges

Despite its vast potential, microscopic-level printing presents its own hurdles:

Color Fidelity: Ensuring true-to-source color rendition at such a granular level can be challenging. Employing top-tier inks, appropriate printing materials, and frequent calibration is paramount.

Balancing Speed and Quality: Rendering minute details takes time. Striking the right balance between swift outputs and impeccable quality is a decision that printers grapple with.

Economic Factors: Precision doesn't come cheap. Premium printers, along with their maintenance and the associated costs of high-grade inks and substrates, demand significant investment.

To overcome these challenges, a mix of adept professionals, continuous learning, and awareness of evolving technologies is essential.

Beyond Visual Appeal

High-resolution printing's applications extend far beyond just attractive visuals. From crafting intricate electronic components to creating detailed medical apparatus and architectural replicas, the capacity to work at a microscopic detail level has transformative implications across sectors.

In Closing

The pursuit of perfection in high-resolution printing is a blend of continuous adaptation, expertise, and the zest to push boundaries. As technology marches on, the horizons of possibilities continue to expand, ushering in a future where the minutiae of designs are replicated with unparalleled accuracy. To be part of this revolution, one needs technical knowledge, a keen eye for detail, and the passion to innovate.

## Tips for maintaining consistency and preventing failures

In the intricate landscape of technological manufacturing, achieving consistency remains a foundational principle. While breakthroughs and innovation are celebrated, the unwavering delivery of those advances truly sets a gold standard. Whether you're overseeing software releases, machine outputs, or any technical task, the mantra is clear: strive for consistency and curtail failures. This article sheds light on strategies to preserve such uniformity.

Recognizing the Imperative of Consistency

Before delving into tactics, it's essential to understand the gravitas of consistency. It's akin to trustworthiness. Within professional realms, consistency bolsters efficiency and fosters stakeholder confidence, be it with customers, collaborators, or associates. Fluctuations or deviations, conversely, can induce operational hiccups, resource wastage, and even safety risks.

1. Comprehensive Record-Keeping

An emphasis on documentation can't be understated. Key elements include:

Standard Operating Procedures (SOPs): Every task should adhere to a well-framed SOP. This serves as the reference point, ensuring uniformity in execution.

Change Records: Document any alterations or updates. Having this trail makes it simpler to trace the origins of any inconsistencies.

2. Routine Training and Skill Updates

Alignment with a workforce is pivotal. Training sessions confirm:

Upgradation of skills to mirror emerging technologies.

Familiarity with protocols, and embedding best practices among the team.

3. Embedded Quality Checks

Infuse quality checks at all stages. Such checkpoints:

Facilitate spot-checks and audits.

Use software test tools for deployment verifications.

Employ checks in production lines.

4. Scheduled Equipment Overhauls

Machinery wear is a frequent consistency nemesis. Routine checks and overhauls:

Spot early signs of equipment fatigue.

Prompt timely interventions, preventing output variance or unexpected breakdowns.

5. Incorporating Automation

While human touch is precious, it can introduce variables. Strategic automation:

Delivers unparalleled precision.

Cuts down on errors in tasks that require repetition.

6. Harnessing the Power of Data

In our modern era, data provides direction:

Utilize analytics to oversee processes. Early anomaly detection can mean quicker solutions.

Predictive tools can spotlight potential issues, facilitating a proactive stance.

7. Promote Transparent Dialogues

Transparent communication is vital. This encompasses:

A platform for team members to share insights or identify challenges.

Continuous interactions across departments, ensuring cohesive actions.

8. Dive-Deep Analysis

When inconsistencies arise, superficial solutions won't cut it. The deep-dive analysis gets to the heart of the issue, paving the way for lasting remedies.

9. Solid Data Backup Mechanisms

In the digital realm, data sanctity is paramount. Robust backups:

Shield against potential data discrepancies.

Provide an assurance of quick recovery from unexpected data challenges.

10. Thorough External Partner Reviews

For those reliant on third-party entities, their performance impacts yours:

Consistently assess the dependability of these external parties.

Draft contracts that emphasize quality benchmarks and consistency metrics.

To Conclude

While consistency might appear routine, it's the silent fulcrum upon which businesses teeter between mediocrity and excellence. It's about impeccable delivery, time after time. Achieving this requires a blend of modern tech tools and a culture centered on diligence and proactive troubleshooting. After all, in the ever-shifting sands of the industry, consistency acts as the steadfast rock, ensuring sustained achievement and growth.

# Chapter Six

## 3D Scanning and Reverse Engineering

### Introduction to 3D scanning technologies

Today's technological environment is continuously evolving, and among its innovations is the capability to accurately digitize the complexities of objects and spaces. This transformative feature is attributed to 3D scanning technologies, reshaping sectors ranging from design and construction to healthcare and media.

Understanding 3D Scanning

Simply put, 3D scanning is the art of harnessing digital data related to an object's form using an array of sensors and light technologies. This data then births a three-dimensional digital representation, often infused with texture and color details.

Diving into 3D Scanning Technologies

There's a broad spectrum of techniques that come under 3D scanning, each having its niche:

Laser Triangulation Scanners: These rely on laser beams and measure the reflection angle when the laser hits an object's surface. Commonly found in tabletop 3D scanners, they're known for their fine detail but can be restricted in their scanning range.

Structured Light Scanners: These scanners project patterns of light onto a subject, and then examine the distortions from a distinct vantage point to deduce depth and contours. A favorite in both personal and professional settings due to its speed.

Time-of-Flight Methods: By measuring the duration a laser beam takes to bounce back after hitting an object, these scanners can measure vast distances, making them ideal for larger projects.

Photogrammetry: This technique stitches together numerous photos taken from varied angles, using software to produce a 3D model. While not always as exact as laser methods, it offers impressive versatility.

Contact-Based Scanning: For certain objects, especially those with intricate interiors, a physical touch is needed. Probes are utilized in contact scanning to gather data points, providing exceptional accuracy.

3D Scanning's Impact Across Industries

The influence of 3D scanning is undeniable and spans several sectors:

Preserving History: Through 3D scanning, delicate historical items and monuments can be digitally archived, ensuring their legacy.

Medical Field: From tailoring prosthetics to mapping bodily structures, 3D scanning has revolutionized personalized treatments.

Media and Entertainment: Realism in modern entertainment mediums owes much to 3D scanning, capturing real-world elements and transforming them into digital assets.

Design and Manufacturing: Engineers often turn to 3D scans of existing items as a reference or foundation for new creations.

Property and Construction: Whether it's creating virtual tours or analyzing existing architectural elements, 3D scanning simplifies and enhances the process.

Navigating the Challenges

Despite its immense promise, 3D scanning presents certain obstacles:

Handling Massive Data: Detailed scans can churn out substantial datasets, which can be challenging to store and handle.

Surface Complications: Surfaces that are shiny, translucent, or exceptionally dark might not always be captured perfectly by some scanning methods.

Budgetary Concerns: Top-tier 3D scanners, especially those tailored for unique industrial tasks, can be a significant outlay.

Yet, even with these challenges, the future of 3D scanning appears bright. With the maturation of software, increased affordability of equipment, and broader industry acceptance, its role in contemporary technology processes is poised to expand.

In conclusion, 3D scanning serves as a testament to the unending quest for knowledge and creation. Acting as a bridge between the physical and digital realms, it unlocks a world of endless exploration and innovation.

## Incorporating scans into design workflows

3D scanning has emerged as a potent tool, bridging the gap between the physical and digital realms. Its incorporation into design workflows provides designers and engineers with a rich tapestry of data, enhancing creativity and precision. Here, we dive deep into effectively weaving 3D scans into your design practices.

Grasping the Unique Advantages

Before intertwining scans into design, it's essential to understand the distinct benefits they bring:

Detailing: Scanners capture objects with unparalleled detail, noting even the tiniest nuances.

Efficiency: Hand measurements or traditional digitization can't compete with the speed of 3D scanning.

Versatility: Be it intricate artifacts or vast landscapes, the scanning spectrum is broad.

Engagement: Once transformed into digital data, these scans can be tweaked, studied, or enhanced in myriad ways.

Strategies for Seamless Fusion

Marrying 3D scanning to design is an art and science combined:

Goal-Setting Before Scanning: Prioritize your objectives. Be it for recreating a design, inspecting quality, or creating a digital repository, your goals shape your scanning approach.

Scanner Selection: Different scanners serve different needs. While some are exceptional at capturing intricate details, others are designed for more expansive scans. Match your project's needs to the right tool.

Data Refinement: Scanning culminates in raw data that often requires refinement. This involves eliminating unwanted data, aligning disparate scans, or cleaning up any irregularities.

Format Transition: Post-refinement, it's crucial to transform the scan into a functional format, ensuring it's conducive to your design software.

Sync with CAD Platforms: Leading CAD platforms usually support 3D scan imports. Once inside, these scans can guide new designs, be juxtaposed with existing elements, or even undergo direct alterations.

Design Revamp: With the 3D scan as your base, design becomes a cyclical process. Designers can tweak, test, and reiterate continuously, eliminating the need to repeatedly refer to the physical model.

Continuous Feedback: Establishing regular feedback channels is key. Rescanning post-design modifications and juxtaposing them with prior designs ensures optimal output.

## Guidelines and Factors to Consider

As you blend 3D scanning with design, several practices can ensure a hitch-free experience:

Managing Voluminous Data: Scans, especially detailed ones, can be data-heavy. Robust storage solutions, both localized and cloud-based, are crucial. Efficient data organization and versioning also come into play here.

Ensuring Software Synergy: Prioritize coherence between the scanner's software, data transformation tools, and your primary design software to prevent potential issues.

Upskilling and Knowledge: Mastery of 3D scanning and its nuances comes with continuous learning. Occasional training ensures your team stays updated and skilled.

Routine Maintenance: Scanners, like any tech tool, demand regular calibration and maintenance to ensure they function at their best.

## The Road Ahead in Scanning and Design

The intersection of 3D scanning and design is only going to get richer with evolving technologies. The combination of augmented reality (AR) and 3D scanning, for instance, can redefine immersive design. As machine learning technologies develop, automated design based on 3D scans could be the next frontier.

To wrap up, the fusion of 3D scanning into design is not a fleeting trend but a paradigm shift. It doesn't just optimize workflows; it propels design into new realms of creativity and precision.

# Practical applications and case studies

In our ever-evolving technological landscape, new methods and tools are continuously reshaping diverse sectors. As these innovations mature, their hands-on applications bring transformative enhancements to industries. In this narrative, we delve into a selection of real-world applications across multiple domains, spotlighting a series of illustrative case studies.

1. Medical Field: Tailored Prosthetics and Implants

Application: One of the most profound impacts of technology in healthcare is the emergence of personalized solutions, particularly in prosthetics and implants. Designed specifically for individual requirements, these innovations have significantly boosted patient comfort and functionality.

Case Study: "Operation Renewal":

In a leading medical institution, a patient with a rare bone condition needed a unique hip joint. Traditional designs were unsuitable. With 3D scanning and printing technology, a custom joint was crafted. After the procedure, the patient experienced enhanced ease of movement and a marked reduction in pain.

2. Archaeology: Digital Archiving and Restoration

Application: Ancient relics and structures, being vulnerable to time, benefit greatly from today's technology, specifically 3D scanning. This approach not only aids in digital archiving but also facilitates potential restorative projects.

Case Study: "Rebirth of an Ancient Metropolis":

When an archaeological expedition discovered an uncharted historical city, they employed drone-assisted 3D scanning to obtain detailed blueprints of the site. This wealth of data facilitated a comprehensive digital reconstruction, offering a window into the city's past. This digital archive ensures the city's stories endure.

3. Vehicle Industry: Efficient Design Prototyping

Application: The process of vehicle prototyping has been revolutionized, shifting from traditionally resource-heavy models to a more efficient system, owing to enhanced simulation techniques.

Case Study: "Revolution Drive's Prototype Evolution":

Revolution Drive, an auto brand, sought to minimize its prototyping timeline. Utilizing state-of-the-art simulation software combined with 3D printing, they tested designs virtually, which trimmed their prototype development phase by nearly half and also substantially cut costs.

4. City Development: Blueprinting Future Cities

Application: To cater to burgeoning urban populations, city developers have turned to modern tools to draft sustainable urban habitats. Advanced simulations provide insights into traffic patterns, resource allocation, and environmental compatibility.

Case Study: "Future City Blueprint":

For an upcoming urban project, city developers combined data analytics with simulation software to model infrastructure designs. Data-driven insights allowed them to pinpoint optimal placements for transportation networks, parks, and utility connections, ensuring long-term resilience and efficiency.

5. Entertainment: Crafting Lifelike Digital Realms

Application: In the realm of entertainment, technology paves the way for the creation of stunningly realistic virtual environments, elevating user engagement.

Case Study: "EpicVista Gaming's Reality Dive":

EpicVista, a trailblazer in digital gaming, leveraged next-gen graphics technology with real-time physics to craft an in-game world that closely mirrored real life. Gamers praised the heightened sense of immersion, lauding the game's realism.

6. Agriculture: Advanced Farming through Data

Application: The dawn of precision farming has seen the integration of data analytics to maximize crop output and resource conservation. Drones, soil monitoring, and analytical tools have revolutionized agricultural decisions.

Case Study: "FarmTech's Data-Driven Agriculture":

In a joint initiative between FarmTech and regional farmers, data from soil sensors and overhead imagery were collected and analyzed. Using this data, actionable strategies were formed, resulting in an uptick in productivity and more efficient resource use.

Conclusion:

The above instances underscore the significant strides made by technological applications in diverse fields. These success stories emphasize the potential of judiciously applied innovations to overhaul industries, introducing solutions that redefine boundaries. The trajectory of technological growth hints at even more exciting prospects ahead.

# Chapter Seven

## Support Structures and Overhang Mastery

### Advanced strategies for minimal supports

In the intricate world of 3D printing, achieving a balance between the structure's integrity and material efficiency is paramount. Support structures are crucial in ensuring printed objects, especially those with complex designs, retain their intended shape during the manufacturing process. But, over-reliance on these can lead to unnecessary material use and post-processing challenges.

To streamline this process, it's vital to embrace techniques that reduce the need for extensive support without compromising the quality of the final print. Here, we delve into strategies tailored to optimize 3D printing processes.

1. Refining Object Orientation

One fundamental way to reduce support structures is by carefully considering the object's placement on the build platform. Slight adjustments in its alignment can drastically cut down the need for support. Aim to position the design so that minimal overhangs face downwards.

2. Adopting Design Principles Geared for Fewer Supports

Implementing Design for Additive Manufacturing (DfAM) concepts can markedly decrease support needs. Leveraging self-sustaining angles, generally below 45 degrees, can negate additional supports. Techniques such as hollowing models, integrating mesh patterns, and adding chamfers or rounded edges also prove beneficial.

3. Leveraging Advanced Software Tools

The software aspect in 3D printing is as significant as the hardware. Modern slicing tools come embedded with refined algorithms that intelligently place supports only where imperative. Additionally, certain software allows users to modify support settings, ensuring a blend of stability and material saving.

4. Tapping into the Potential of Soluble Supports

Soluble supports present a groundbreaking approach. Crafted from materials that dissolve, these supports can be seamlessly removed after printing by immersing the object in specific solutions. This not only facilitates complex designs but also streamlines post-processing.

5. Shifting to Innovative Support Structures

Instead of the conventional blocky supports, consider the more efficient tree-like supports. They branch out to support different parts of the design, using less material. Conical supports, broader at the bottom and tapering at the top, are another option that combines sturdiness with material conservation.

6. Strategies for Reducing Gaps

Managing minute gaps in designs can be tricky when minimizing supports. By using dynamic bridging methods, it's possible to cover these spaces directly with the print material, thus sidelining supports. Adjustments in printing speed, temperature, and cooling can aid in ensuring these bridges maintain their shape.

7. Adopting a Test-and-Refine Approach

Having insights into the printer's unique capabilities is vital. By repeatedly printing challenging sections of a design, one can gather key learnings on how to optimize supports. This hands-on approach can lead to tailored strategies for each specific printer.

8. Trying Out External Stability Techniques

In some instances, external measures can substitute built-in supports. Methods include maintaining a steady temperature in the printer's environment, minimizing external vibrations, or even using manual support techniques for smaller prints.

Conclusion

Navigating the complexities of 3D printing requires a blend of art and science. While the utility of support structures can't be denied, the aim should always be to maximize efficiency and minimize waste. By integrating the aforementioned strategies and capitalizing on the advances in technology, one can inch closer to achieving this delicate balance.

## Designing for self-supporting structures

In the intricate world of 3D printing and digital design, striking a balance between resource conservation, functionality, and visual allure stands at the forefront. A design that stands on its own without the need for auxiliary aids, or only a minimal set of supports, is a testament to the power of modern additive manufacturing. This delve into self-sustaining design touches on the key components, from geometry to printing dynamics, that contribute to creating a fully supported structure.

Central to this discussion is the role of geometry. Overhanging features, which jut out without direct support underneath, pose challenges in 3D printing. While typical Fused Deposition Modeling (FDM) printers can tackle up to 45-degree overhangs, transcending this limit calls for deep material insights, printer acumen, and inventive strategies.

To avoid additional support, certain geometric strategies can be employed:

Tapered designs: By narrowing structures at their uppermost part, supports can be eliminated. This gentle slope ensures each layer gains adequate backing from its predecessor.

Spanning bridges: When a design has two elevated pillars, it's feasible to stretch material between them, constructing a 'bridge'. Mastering this technique requires understanding the interplay between printing speed, extrusion amount, and rapid cooling.

Leveraging arches and curves: The inherent sturdiness of these shapes can be utilized for designing components that are both visually engaging and self-standing.

It's also essential to factor in the behavior of materials during printing. Different substrates have unique characteristics that affect their flow and how they solidify. Recognizing these properties can refine design choices.

Printer dynamics also play an integral role. A reduced layer height, for instance, can better handle overhanging sections, as smaller layers provide improved support. Adjusting extrusion dimensions, particularly for the outermost layers, strengthens potential weak spots.

Modern 3D printing software has also introduced features that automatically adjust settings based on design complexity. This dynamic adjustment, often called 'adaptive slicing,' fine-tunes layer dimensions based on design needs. Utilizing this can drastically improve the print quality of sophisticated, self-supporting designs.

Another influential method is 'topological optimization.' Here, software identifies and manages stress areas in a design. It repositions materials—taking from low-stress zones and reinforcing high-stress points. The result is a structure that's not only self-supporting but also efficient in its material use.

Finally, after printing, certain treatments can enhance a design's resilience. Controlled heat cycles, for instance, can alleviate built-up internal stress, and improve the bond between layers. Some modern materials even offer the advantage of post-production annealing to bolster part durability.

To wrap up, as the domain of additive manufacturing evolves, there's an increasing emphasis on designs that are optimized for self-support. By meshing geometry, material knowledge, printer mechanics, and the latest software tools, designers can navigate the path from concept to tangible, self-standing structures more seamlessly.

## Innovative materials for dissolvable supports

Additive manufacturing, often known as 3D printing, has undergone a transformative shift over the recent years. One of the standout innovations is not just in the realm of printer capabilities or design software, but specifically in the domain of materials designed for dissolvable supports.

During the 3D printing process, many designs often have parts that extend or have intricate shapes requiring provisional support. Traditional supports can be challenging to remove and might mar the finished product. Enter dissolvable supports – the game-changer that streamlines post-production, delivering refined results.

Let's explore some of the notable materials introduced for this application:

Polyvinyl Alcohol (PVA):

A frontrunner in the world of dissolvable supports, PVA is typically used with Fused Deposition Modeling (FDM) devices. Easily dissolving in water, PVA is the go-to for models that need post-production immersion. It shines particularly when working on designs with hard-to-reach cavities.

High Impact Polystyrene (HIPS):

Paired with ABS (Acrylonitrile Butadiene Styrene) in a dual-material setup, HIPS dissolves in d-Limonene, a citrus-derived solvent. After printing, the model's immersion in d-Limonene ensures the HIPS supports vanish, revealing the untouched ABS model.

BVOH (Butenediol Vinyl Alcohol Copolymer):

Emerging more recently in the dissolvable supports scene, BVOH boasts of faster water dissolution rates than PVA. Its adaptability is commendable, fitting seamlessly into multi-material print tasks. Enhanced adhesion qualities of BVOH ensure seamless transition zones between the model and its supports.

Polyethylene Glycol (PEG):

Already well-established in various industries, PEG's solubility in water makes it a contender in 3D printing. Its biocompatibility is a boon for medical applications that might require temporary scaffolds.

The benefits of these materials in the additive manufacturing landscape are multifold:

Refined Surfaces: Conventional supports might compromise a print's surface where it attaches. In contrast, dissolvable variants ensure a seamless finish, both in appearance and structure.

Empowered Designing: Designs that lean on the intricate side can be risky with standard supports. The dissolvable variants encourage bolder design choices without the associated risks.

Enhanced Efficiency: Removing traditional supports can be a lengthy process. Dissolvable ones expedite post-processing, aligning with swift project completions.

Diminished Risks: The primary model can sometimes be jeopardized during traditional support removal. Dissolvable alternatives eliminate such contingencies.

While these materials are groundbreaking, it's vital to ensure their alignment with the primary materials used, printer specifics, and post-processing procedures. For instance, pairing PVA with high-temperature materials may not be ideal, given PVA's temperature sensitivity.

In summation, the introduction of such avant-garde dissolvable support materials marks a significant stride in additive manufacturing. They offer designers and manufacturers the latitude to produce detailed and high-fidelity prints without the traditional support-removal hindrances. As the material science behind 3D printing evolves, the horizons of what's possible in this field continue to expand, pushing the boundaries of creativity and application.

# Chapter Eight

## Complex Assemblies and Moving Parts

### Designing interlocking parts

The art of developing components that interlock seamlessly is deeply rooted in both technical expertise and innovative thinking. Throughout the design and engineering landscape, crafting interlocking elements that lock together without external attachments or glues is paramount. This intricate journey of designing these elements requires a solid grasp of foundational principles, material behavior, and the end purpose they serve.

The Bigger Picture:

From time immemorial, interlocking designs have been an integral part of structures around us. Whether it's the child's puzzle piece or a grand architectural piece, these designs offer more than just stability. They pave the way for material efficiency and elevate design aesthetics by eliminating external fastening elements.

1. Beginning with an Idea:

The initial phase of designing interlocking parts revolves around visualizing the end product, understanding the directions in which forces will operate, and the interplay between components. Using tools like rough drafts, digital design sketches, or advanced 3D software can streamline this process.

2. Accuracy and Precision:

Interlocking mechanisms demand the utmost precision. A slight misalignment can compromise the integrity of the mechanism. It's imperative for designers to align with manufacturing capabilities and be aware of how different materials might behave to ensure perfect fitment.

3. Material Choices and their Behavior:

Materials respond differently under various conditions. For instance, metals provide solidity, whereas plastics might give the necessary elasticity for mechanisms like snap fits. Considering potential material reactions, including expansion or shrinkage due to temperature fluctuations, is key.

4. Popular Interlocking Designs:

Dovetail Designs: Predominantly observed in woodcraft, these have trapezoid-shaped extensions fitting perfectly into corresponding slots.

Tongue and Groove Method: This involves a protruding element (tongue) fitting into an indent (groove).

Snap Fits: Common in plastic designs, these use material elasticity to hold parts together.

Bayonet Mechanisms: A rounded method where insertion is followed by a twist to lock.

5. Longevity Concerns:

Especially with mechanisms that are frequently used or moved, ensuring their long-term functionality is essential. Factoring in aspects like environmental influences, wear and tear, and force dynamics can offer insights.

6. Visual and User Comfort Considerations:

While functionality is the mainstay, the visual appeal can't be sidelined. Flawless transitions, concealed connections, and refined design details enhance the visual appeal. If the design involves human touchpoints, user comfort should be prioritized.

7. From Concept to Reality:

While theoretical designs provide a starting point, real-world testing ensures functionality. Using techniques like 3D modeling or swift prototyping gives tangible models for trials. Feedback from these tests can be looped back into refining the design.

8. Future-Proofing the Design:

It might seem paradoxical, but thinking about how the interlocking parts can be separated is as vital as designing their union. In sectors such as electronics, this foresight allows for updates or maintenance.

9. Unifying and Modular Approach:

For designs that might belong to a broader system or have subsequent versions, it's wise to maintain a level of standardization. This promotes a modular approach, ensuring adaptability and fostering user trust.

10. Considering the Broader Impact:

In today's ecologically aware age, there's a tilt towards designs that reduce reliance on harmful glues or fasteners. Eco-friendly designs, while being sustainable, also present long-term economic advantages.

In a nutshell, creating interlocking components is a sophisticated blend of technical acumen, creativity, and aesthetic sense. As the tools of the trade, like simulation software and comprehensive CAD platforms, evolve, professionals are better poised to create groundbreaking and efficient interlocking designs. Beyond their simple appearance, these mechanisms have transformative potential across multiple realms of product design and end-user interaction.

## Print-in-place techniques for movable components

The landscape of 3D printing has witnessed a significant shift with the advent of creating components that are operational right after printing, eliminating the requirement for subsequent assembly. Termed as the print-in-place methodology, it streamlines the production of articulated items, marrying efficiency with ingenuity. By delving deep into the nuances of this method, one can harness the comprehensive capabilities of 3D printing, crafting movable elements that resonate with functionality and novelty.

Grasping Print-in-Place:

Print-in-place is distinguished by its elimination of post-print assembly. This is actualized by fabricating movable elements in their conclusive position, making sure there are intervals or gaps that facilitate movement upon completion. This method shines in its versatility, from formulating pliable joints to sophisticated gear mechanisms.

1. Factors before Designing:

Successful execution of print-in-place pivots on certain anticipatory design aspects. Recognizing the optimal print orientation can be a game-changer. It's essential to ensure that mobile components are situated in a manner that they remain unobtrusive to the print head and are safeguarded from unintentional fusion due to inadequate gaps.

2. Space and Precision Levels:

The efficacy of print-in-place components is often contingent on the specific intervals between them. Although the exact space can oscillate based on printer specifics, filament choice, and design sophistication, an initial gap of 0.5mm often proves beneficial, but hands-on experimentation is advised for fine-tuning.

3. Material Choices:

Diverse 3D printing substrates exhibit variations in suppleness, durability, and post-fabrication contraction. For applications that demand some stretch, like flexible connections, materials like TPU or nylon excel. In contrast, sturdier materials such as PLA or ABS may be more apt for interconnecting gears.

4. Navigating Gravitational Pull:

Designing parts that extend, like protruding arms, brings the potential risk of sagging due to gravitational forces. Introducing temporary reinforcements or a slight upward inclination in design can mitigate this.

5. Infusing Joints and Pivots:

Incorporating joints, especially in elements like mechanized arms, is transformative. Typically, a pivot mechanism is central to the hinge's movement. Ensuring ample spacing and leveraging flexible substrates can amplify its effectiveness.

6. Gear Configurations:

When gears are the focal point in print-in-place designs, their alignment and interspacing take precedence. Beyond standard space evaluations, the contour of gear cogs and their interlinking should be attentively fashioned for fluid rotation.

7. Test and Refine:

Owing to the intricate nature of print-in-place, repeated testing isn't just a suggestion but a necessity. Continuous iterations based on tangible outcomes lead to designs that align with the envisioned functionality.

8. Fine-Tuning and Upkeep:

Even though print-in-place aims to minimize aftercare, some follow-up actions, like extraction of provisional supports or smoothening edges, might be warranted. Also, routine maintenance can bolster the longevity of the components.

9. Pushing Boundaries:

With the evolution of diverse material printers and advanced software, the horizon for print-in-place is expanding. Crafting with varied materials in a singular entity can introduce multiple rigidities within the same construct.

10. Exploring New Frontiers:

The scope of print-in-place isn't confined to miniature models. As technology progresses, printing large-scale dynamic structures, ranging from construction elements to extensive machinery is coming to the forefront.

To wrap up, the realm of print-in-place for movable components stands as a groundbreaking segment within 3D printing. Meticulous planning coupled with tech advancements offers the prospect of generating intricate, agile, and utilitarian components sans the assembly phase, redefining the fabrication paradigm. With relentless progress in tools and methodologies, print-in-place is poised to catalyze innovations, reshaping industry standards.

## Tips for ensuring free movement post-print

Crafting functional 3D prints with moving parts requires attention to detail from design to post-processing. The ability to achieve free movement after printing is a testament to both the printer's capabilities and the designer's expertise. Here's a compilation of insights to guide you in achieving optimal mobility in your 3D projects.

1. Importance of Proper Calibration:

Before you embark on any intricate printing project, make sure your 3D printer is correctly calibrated. A machine in its prime state ensures the faithful reproduction of designs, leading to parts that interact as envisioned.

2. The Layer Resolution Game:

For prints with complex moving elements, opting for finer layer resolutions is wise. While it may lengthen the print time, layers between 50 to 100 microns typically result in smoother finishes, leading to lesser friction between moving parts.

3. Always Ensure Enough Clearance:

When designing, always factor in some space between components. Although 3D printers are accurate, materials can sometimes exhibit minor expansions. A safe bet is maintaining a clearance of 0.5mm, adjusting based on material and machine specifics.

4. Master the Art of Bridging:

Bridging refers to printing horizontal stretches without supports. By perfecting the settings associated with bridging, like optimal cooling and correct print speed, you ensure components remain true to their design.

5. Dynamic Support Structures:

When parts are elevated during the printing process, removable supports can be invaluable. However, it's crucial to design these supports in a manner that ensures they don't hinder the final movement of the print.

6. Material Choices Matter:

Different filaments offer varied properties. Materials like TPU or nylon are more flexible than traditional ones like PLA. A deep understanding of your filament's characteristics can influence design decisions for movement.

7. Post-Print Tweaks:

Sometimes, minor adjustments after printing, such as careful sanding or lubricating, can make parts more mobile. Proceed delicately, ensuring the essence of the design remains intact.

8. Overlapping Design Strategies:

For pieces like balls and sockets that are supposed to interact closely, design with a touch of overlap. This slight fusion during printing ensures a snug fit while maintaining movement.

9. Find the Temperature Sweet Spot:

Too hot, and your material might flow too freely, risking fused parts. Too cold, and layer adhesion might suffer. Determine the right temperature for both material and design.

10. Factor in Tolerances:

By designing with possible discrepancies in mind, you can avoid unexpected obstructions or fusions in the final print.

11. Consider Modular Approaches:

For multifaceted designs, printing parts separately and then assembling them can ensure each piece functions optimally.

12. Regulate Cooling Patterns:

Avoid rapid cooling, which can result in warping. A consistent and controlled cooling approach is recommended for prints with tight spaces or intricate details.

13. The Value of Prototyping:

Before going all in, print a smaller version or just the components in question. This preliminary step can help refine your process, saving time and materials.

14. Introduce Snap-Off Points:

If you anticipate parts might adhere together, add a designated breakpoint that can be easily separated after printing.

15. Stay in the Loop:

The world of 3D printing is continually advancing. Regularly updating your knowledge can expose you to refined techniques and tools.

In conclusion, producing a 3D print with fluid post-print movement combines equipment readiness, intricate design considerations, and the apt choice of materials and settings. Adopting a blend of these recommendations will ensure your 3D-printed objects replicate the freedom of movement seen in their digital designs.

# Chapter Nine

## Surface Finish and Advanced Post-processing

### Techniques for achieving mirror finishes

Attaining a mirror finish—a surface that gleams with shine and clarity—combines both technical skill and nuanced understanding. Such finishes not only enhance aesthetics but also bolster protection for the underlying material. This discussion unveils the diverse approaches available for producing these pristine finishes, from metals to polymers.

1. Metals and Their Counterparts:

Surface Smoothing:

Start with an abrasive tool, like coarse sandpaper, to even out irregularities. As the process advances, switch to finer abrasives and conclude with specific polishing agents.

Chemical Polishing:

Particularly effective for stainless steel, this method chemically refines the metal's surface. The metal serves as the anode, undergoing a process where elevated regions dissolve more rapidly than depressions, ensuring uniformity.

Shine Enhancement:

This final phase uses a fabric wheel alongside polishing agents. The spinning wheel, the pressure, and the compound's abrasiveness synergize to grant the desired shine.

2. Polymer Materials:

Solvent-Based Smoothening:

For materials like polycarbonate, solvent vapors are used. The vapors momentarily soften the surface, and upon solidifying, the facade becomes both glossier and clearer.

Controlled Melting:

A purposefully managed flame, preferably from a specialized torch, is swept across the polymer, which briefly softens the surface. As it cools, a shiny exterior emerges, but mastery is required to avert damage.

Abrasive Enhancement:

Polymers, like metals, benefit from abrasives. Nevertheless, one should be cautious to prevent heat build-up, which might alter the material.

3. Glassy and Ceramic Materials:

Refinement of Ceramics:

The fragile nature of ceramics often necessitates the use of diamond-coated tools. Initial stages involve substantial grinding, transitioning to finer techniques for the polished outcome.

Optical Glass Refinement:

Precision is paramount, especially for optical glasses. Specialized pads combined with specific slurries yield a reflective finish.

4. Protective Layers:

Transparent Shields:

Apart from adding a glossy layer, clear coatings or lacquers defend the base material. Once settled, they can be further refined to elevate their reflective attributes.

Shiny Sprays:

For non-metallic bases, sprays with chrome or metallic properties followed by a clear shield can replicate a mirror finish. Consistency in application is crucial.

5. Modern Techniques:

Laser-Driven Refinement:

Lasers offer a swift and potentially automated method but require finesse to prevent unwanted heat damage or alterations.

Atomic-Level Smoothing:

Common in specialized industries like semiconductors, ion beams achieve remarkable smoothness by shedding the topmost layer, albeit at a higher cost.

Pro Tips for the Perfect Shine:

Starting Right: Preparation is pivotal. Prior stages like sanding should be done diligently, as overlooked defects can compromise the final appearance.

Uniform Application: For hand-driven methods, a steady hand ensures no unwanted patterns mar the surface.

Maintain Purity: Minuscule contaminants can blemish the finish. A pristine environment and regular cleaning can mitigate this.

Prioritize Safety: Especially when chemicals or fast-moving equipment are involved, safety equipment and guidelines must be adhered to.

In essence, delivering a mirror finish demands a comprehensive grasp of the chosen material and the techniques at hand. With informed choices and precision in execution, the outcome can be both visually striking and enduring.

## Advanced painting, sealing, and coating methods

The realms of painting, sealing, and coating have witnessed considerable evolutions, courtesy of breakthroughs in technology. Current strategies and substances offer strengthened resilience, amplified aesthetic appeal, and tailored functional properties. Diving into this space, we encounter an intriguing blend of scientific disciplines.

1. Electrocoating:

Relying on the dynamics of charged entities, this technique ionizes paint with an opposing charge to the material, ensuring a magnetic attraction for an even layer. The method stands out for its minimal wastage, maximized transfer rate, and impeccable finish.

2. Low Pressure, High Volume (LPHV) Mechanisms:

LPHV systems leverage reduced air pressures for precise application and minimal spray wastage. Here, the emphasis is on volume over pressure, facilitating controlled and effective layering.

3. Thermal Plasma Coating:

A favorite in sectors like automotive and aerospace, this technique pushes powder through a super-heated plasma stream, leading to its deposition on a material. The outcome is a tight, cohesive layer, suitable for electrically conductive or wear-resistant applications.

4. Monomolecular Layer Deposition (MLD):

This method focuses on introducing precursors sequentially for reactions with the base material. Layer upon layer, the structure is built, allowing for exceptional control over thickness. MLD finds its niche in nano-technological and microelectronic applications.

5. Micro-coatings:

By integrating nanoparticles, these coatings offer enhanced properties compared to conventional ones. They can impart unique attributes like water repellence, resistance to scratches, or self-repairing capabilities.

6. Vacuumed Metal Deposition:

Commonly employed for decorative finishes or lighting applications, this process involves metal vaporization in a vacuum space, which subsequently settles on a base material. Apart from its striking look, it can act as a protective shield or enhance conductivity.

7. Light-Activated Curing:

Differing from traditional drying or curing, these coatings rely on UV rays to start a swift polymerization process. This results in a near-instantaneous solid surface, ideal for quick-paced production lines.

8. Polymers with Shape Adaptability:

Redefining sealing solutions, these polymers morph their shape when subjected to stimuli like UV radiation or heat. This feature allows for the development of adaptable structures or self-mending seals.

9. Thermal Insulating Layers:

Indispensable for components subjected to elevated temperatures, such as those in jet engines, these coatings offer a thermally protective layer. Compounds like yttria-stabilized zirconia are popular for their stability and thermal resistance.

10. Anti-Corrosion Paints:

These modern paints serve more than just a visual purpose; they combat corrosion. Incorporating specific compounds or nanoparticles, they form a shield against environmental threats, extending the longevity of the material beneath.

Recommendations for Advanced Application:

Substrate Readiness: Regardless of the technique employed, ensuring the base material is primed and ready is paramount. This includes processes like cleaning, smoothing, and occasionally, priming.

Ecological Impacts: Some coatings might come with environmental implications during production or disposal. It's crucial to assess their ecological impact.

Safety First: The use of certain advanced methods might come with risks due to chemicals or machinery. Adhering to safety guidelines and utilizing protective equipment is essential.

Aftercare: Even top-tier coatings might need periodic maintenance. Following care instructions, whether it’s safeguarding from UV exposure or maintaining specific temperatures, ensures the coating remains at its peak.

In a nutshell, the domain of painting, sealing, and coating is brimming with state-of-the-art techniques that cater to the nuanced demands of today's industries and design preferences. While they are rooted in time-tested scientific principles, their modern adaptations address the sophisticated requirements of the present-day world.

## Joining and welding 3D printed parts

The vast world of 3D printing often demands the assembly of smaller segments to form larger constructs, primarily due to printer size limitations or the convenience of transportation and assembly. This step isn't as simple as combining parts; it requires a deep knowledge of the materials involved and the best methods to merge them.

Deciphering Material Characteristics:

Prior to deciding on a joining technique, it's paramount to understand the specific characteristics of the 3D printed material. For instance, thermoplastics can be remelted and fused, which makes them compatible with thermal welding. Metals, conversely, might necessitate more classic welding approaches, adjusted for the specific metal and printing method utilized.

Heat-Based Fusion:

This approach is primarily for thermoplastic materials, utilizing controlled heat. By warming the joining areas to a semi-fluid state and then uniting them, a connection nearly as robust as the original material can be realized. Achieving the correct temperature is crucial: excessive heat may warp the parts, whereas too little can lead to a feeble bond.

Welding Through Friction:

In this method, the heat generated from friction between components is used for fusion. This is especially beneficial for materials sensitive to direct heat. Spin welding, a subset of this method rotates one component at high velocities against a stationary one.

Using Adhesives:

Occasionally, the ideal way to merge two parts is through adhesive bonding. Using substances like epoxy or superglue can provide a sturdy bond. The success of this method often rests on choosing an adhesive that's synergistic with the 3D printed material and ensuring the surfaces are prepped and clean.

Welding with Ultrasonics:

This method employs ultrasonic sound waves to produce quality welds, particularly with thermoplastic materials. The sound waves generate heat at the junction, leading to a bond. Its efficiency and quick processing time render it favorable in many applications.

Welding Metals:

For 3D-printed metallic components, traditional methods such as TIG or MIG welding are applicable. But, given that 3D-printed metals may have unique microstructures, the welding parameters might need fine-tuning.

Magnet-Based Assembly:

Incorporating magnets during the printing process can simplify post-production assembly. While this doesn't provide the strength of other methods, it's versatile, especially in designs that are modular.

Mechanical Connections:

By designing parts with features like snap fits or interlocking joints, you can bypass many external joining methods. The effectiveness of these designs depends on the chosen material and design details.

Post-Assembly Enhancements:

Post-joining, some refining might be required to perfect both the functionality and appearance. This could include smoothing out the fusion lines, filling any remaining gaps, or coating the area to achieve a seamless look.

To summarize, as 3D printing's role in industries like manufacturing expands, mastering the art of joining and welding these components becomes paramount. The method of choice will be influenced by various factors including material type, final application, and desired aesthetics. As technology progresses, these methods will undoubtedly evolve, yielding more resilient and visually pleasing results.

# Chapter Ten

## Functional Prototyping and Testing

### From design to functional prototype

The transformation from a mere idea to a tangible, functional prototype is an essential phase in the field of product creation and manufacturing. This change isn't just the act of making thoughts material; it's about converting conceptual notions into actionable, concrete entities. This process is a delicate balance of art, technology, and continuous adjustments. Here, we navigate through this captivating progression, detailing the essential steps and elements that carry a concept through to a working prototype.

1. Ideation: The Starting Point:

All products start as a spark of inspiration. Whether driven by market demands, innovative solutions to persistent challenges, or pure creativity, this initial stage lays the groundwork. At this point, cross-functional teams, including innovators, designers, and decision-makers, join forces to ideate, conceptualize, and outline the project's end goal.

2. Initial Sketching: Visualizing the Idea:

Once the core concept is defined, the next step involves breathing life into it. Designers use various tools and techniques to produce the product's initial visual representations. These rough outlines help set expectations regarding the product's look, size, and basic functionalities.

3. Digital Drafting: The 3D Transition:

With foundational sketches ready, the conceptual design enters the digital arena. Sophisticated CAD (Computer-Aided Design) platforms allow for intricate 3D modeling. This step provides a detailed look at the product, making it possible to explore its dimensions, pinpoint potential design challenges, and offer diverse perspectives.

4. Viability Check: Matching Vision with Reality:

Before advancing, it's imperative to ensure the design's real-world feasibility. Multiple sectors, ranging from technical to economic, scrutinize the design's practicability. They evaluate aspects like material compatibility, cost factors, production limitations, and overall usability, making adjustments as necessary.

5. Tweaking: Pursuit of Excellence:

Feedback from the feasibility review informs further adjustments to the design. This stage is all about optimization, tackling limitations, enhancing its user-centric features, and aligning it with production realities.

6. Material Choices: Picking the Right Foundations:

Materials play a pivotal role. They determine factors like product resilience, appearance, heft, and cost-effectiveness. The intended function of the prototype, its anticipated use cases, and the environment it will operate guide material decisions.

7. Quick Prototyping: A Touch of Reality:

With the design polished and materials chosen, the first real-world version emerges via quick prototyping. Using methods such as 3D printing, CNC operations, or artisan craftsmanship, a physical model is crafted. This model, though preliminary, is vital for evaluations and assessments.

8. Real-world Evaluation: Stress Testing:

Even in its nascent form, the prototype must demonstrate its functionality. It is subjected to a series of tests that mimic its intended environment, determining its robustness, performance, user comfort, and other factors. These evaluations offer crucial insights for subsequent modifications.

9. Integrating Feedback: Collective Wisdom at Play:

Reactions and inputs from potential users, collaborators, or internal teams are crucial at this juncture. By incorporating these perspectives along with insights from real-world testing, the prototype undergoes further finetuning.

10. The Final Form: A Masterpiece Emerges:

With all the feedback, tweaks, and evaluations in place, the end result – the final prototype – is crafted. This piece closely mirrors the eventual product, showcasing its aesthetics and practicality.

In Closing:

The progression from a mere idea to a functioning prototype is a blend of creativity, cutting-edge technology, and perpetual enhancements. It calls for interdisciplinary teamwork, the adept use of innovative tools, and an uncompromising attention to detail. The end product, a fully realized prototype, stands as a monumental achievement in the product development lifecycle.

## Stress testing and validation for 3D printed parts

As 3D printing takes center stage in modern manufacturing processes, understanding the resilience and efficacy of printed parts becomes critical, especially in sectors like aerospace, automotive, and medical devices. Thorough testing and validation are fundamental in affirming the functionality and safety of these components.

1. Why Stress Test?:

3D printed items often deviate from traditionally manufactured ones, especially in their microstructures and resultant properties. Subjecting parts to conditions exceeding typical usage is necessary to determine their breaking limits, validate designs, and assess the performance of the materials.

2. Diverse Stress Tests for 3D Printouts:

Tensile Assessments: By exerting a singular force, this method measures a component's resistance to being pulled apart, establishing its ultimate tensile strength and the extent of elongation before breaking.

Compression Examinations: This test reviews how a part reacts to forces pushing it together, highlighting potential buckling and the overall compressive strength.

Endurance Tests: Parts in real-life scenarios often face fluctuating loads. Cycling through various load magnitudes can determine the component's lifespan and identify potential fatigue-related failures.

Shock Analysis: This measures the component's durability against sudden, unexpected forces.

Thermal Evaluations: Depending on their composition, 3D-printed components might encounter a range of operational temperatures. Assessing their reactions to temperature fluctuations is essential.

3. Material Factors in Stress Tests:

Each 3D printing material offers unique stress test results. For instance, a component crafted from PLA may behave differently under testing than one made of metal or ceramic. It's essential to consider the part's function when selecting materials and to account for any post-processing, which might change material characteristics.

4. Verifying 3D Printed Items:

After stress testing, a rigorous validation process confirms the part's consistent functionality.

Size and Shape Verification: Using advanced tools like laser scanners, one can ensure the 3D printed part aligns with its design specifications.

Evaluating Surface Quality: Many parts have strict surface finish requirements that affect various attributes. Tools can assess this finish to ensure it's up to par.

Density and Void Checks: This is paramount for metallic 3D prints, ensuring the integrity of the component.

5. Simulating Stresses:

Before physical prototyping, tools like Finite Element Analysis (FEA) offer a glimpse into how the part might behave under stress. This step can lead to early design modifications, making the process more efficient.

6. Industry-Specific Testing:

Depending on the field, there might be unique testing requirements, such as checking for biocompatibility in medical devices or running fire resistance tests for aerospace parts.

7. Prototype, Test, Repeat:

It's common for initial designs to fall short of performance expectations. Combining rapid prototyping with immediate testing facilitates quick design adjustments, optimizing the development process.

Conclusion:

Ensuring the safety and performance of 3D-printed components requires thorough stress testing and validation. As the world becomes more reliant on 3D printing technologies, industry-specific testing standards will be paramount. Committing to these evaluations will bolster confidence in additive manufacturing outputs.

## Iterating and refining based on feedback

The cyclical journey of creation, from ideation to realization, frequently revolves around feedback. Whether in product design or engineering, this rhythm of design, testing, review, and refinement is integral. Such feedback-centric methodologies elevate quality and ensure alignment with the intended user's expectations and necessities.

The Power of Feedback in Design Iterations

Feedback uncovers hidden opportunities and challenges that might have been overlooked by designers or engineers. When feedback is incorporated, it translates into solutions that are better tailored to meet real-world needs, enhancing both efficacy and functionality.

1. Categories of Feedback

Specialist Feedback: These insights come from those deeply embedded in the field, possessing a wealth of technical expertise. Their perspectives can pinpoint essential design alterations or functional nuances.

End-user Feedback: This originates from those who experience the product in everyday situations. Such feedback is crucial to gauge the product's intuitiveness and to identify unforeseen challenges.

Automated Responses: Emerging technologies now enable systems to furnish feedback based on pre-set parameters or simulated outcomes.

2. Iterative Stages Characterized by Feedback

Ideation Iteration: In these foundational stages, feedback can lead to significant directional shifts, refining the core idea.

Prototype Iteration: With a working model in place, feedback becomes more specific, touching on aspects like usability, aesthetics, and practicality.

Refinement Iteration: As the product nears completion, tweaks in this phase respond to detailed testing or preliminary market reactions, ensuring the product's robustness.

3. Methods to Garner Effective Feedback

Feedback Forms and Polls: Beneficial for acquiring numerical data, assessing overall user satisfaction, or pinpointing prevalent concerns.

Discussion Panels: Provide an in-depth exploration of certain facets, yielding rich qualitative data and fostering open dialogues.

User Interaction Analysis: Observing user interactions with a product offers impartial insights into its merits and areas of improvement.

4. Turning Feedback into Constructive Changes

Simply amassing feedback isn't the endgame. A systematic interpretation is pivotal to extracting valuable takeaways.

Identifying Trends: Recognize issues or suggestions that emerge consistently. These recurrent topics usually signal primary areas of enhancement.

Ranking Feedback: Not all feedback carries equal weight. Identify which modifications will most profoundly influence user experience or functionality.

Assessing Implementation: All feedback, while valuable, might not be practical. Before initiating changes, evaluate the technical, financial, or time-related implications.

5. The Process of Refinement

Post-feedback analysis, the next step involves meticulous refining. This can range from subtle modifications to comprehensive redesigns.

Team-based Refining: Engage diverse teams in this process. A collaborative approach, combining the insights of engineers, designers, and other key players, ensures comprehensive problem-solving.

Swift Model Testing: Contemporary tools allow for rapid modeling and testing of changes, fostering an agile iteration loop.

Chronicle the Changes: Keep a comprehensive record of all iterations, the feedback catalysts, and the rationale behind each decision. This archive not only aids current projects but also informs future ones.

In Summary

Feedback-oriented design refinement is a perpetual journey. The commitment to refining based on post-debut feedback might birth product upgrades, enhanced functionalities, or entirely revamped offerings. In a dynamic commercial landscape, the ability to heed, adapt, and enhance is not merely commendable but vital for sustained relevance.

# Chapter Eleven

## Quality Control and Troubleshooting

### Techniques for consistent high-quality prints

Securing unwavering quality across print outputs marks the expertise of a true digital printing aficionado. Whether engaged in 3D printing or the more conventional 2D printing methodologies, it's the uniformity in output that establishes reliability and earns stakeholder trust. While equipment and technological advancements are undeniably significant, a deep grasp of specific nuances and an adept handling of mutable factors remain central. Delve into these crucial techniques and insights that assure top-notch print quality every single time.

1. Grasping Material Dynamics and Their Treatment

The canvas dictates the art's potential.

Choosing Wisely: Picking the right filament, paper, or ink can make or break the print output. Prioritize high-quality, tried-and-tested materials tailored for the task at hand.

Safekeeping: Age and environmental factors can compromise material integrity. Shield them from dampness or sunlight, often in tandem with moisture-absorbing agents.

2. Precise Calibration and Timely Upkeep

A tool's efficiency mirrors its well-being.

Regular Assessments: Frequent equipment evaluations can preempt potential issues, promoting timely interventions and replacements.

Calibration Rituals: Periodic printer calibrations maintain operational accuracy. This encompasses balancing the 3D printer bed or ensuring accurate color representations in digital prints.

3. Tailoring Print Parameters

Crafting masterpieces demands mastery over the tool.

Determining Layer Thickness and Clarity: For 3D prints, the layer's thickness can influence the finish. Lean towards finer layers for intricate outputs, recognizing the trade-off with time. For 2D, DPI configurations influence image crispness.

Speed Considerations: Expedited outputs can sometimes be at odds with quality. Balance speed against the specific printer's capabilities and material specifications.

Optimal Temperature Regulation: For 3D endeavors, both the bed and nozzle temperatures are pivotal. In 2D scenarios, a pristine and optimally heated print head is fundamental.

4. Initiating with Foresight

Quality prints owe much to a thoughtful start.

Scouring Designs: Prior to initiating the print, diligently review designs for potential inconsistencies or anomalies.

Slicing Tools: In 3D paradigms, slicing tools transform models into printable layers. Opt for comprehensive software, offering granularity in settings and requisite support frameworks.

Sampling: For intricate projects, preliminary test runs can unearth unforeseen challenges, ultimately conserving valuable resources.

5. Post-Print Nuances

Quality doesn't end when printing does.

Gradual Cooling: Hastened cooling can skew results, inducing deformities or color disparities. Using heated beds in 3D prints can mitigate such issues.

Refinements: Meticulously remove any superfluous elements or residual materials. For 2D outputs, ensure ink settling and mull over protective enhancements like lamination.

Assess and Document: Post completion, subject the output to scrutiny for potential aberrations. Chronicle these observations for progressive refinement.

6. Cultivating an Ideal Ambient Setting

The surroundings weave into the print's fabric.

Consistent Ambient Temperatures: Temperature variances can skew 3D print outcomes due to material behaviors.

Moisture Management: Elevated humidity can tarnish materials and interfere with seamless printing.

Maintaining Purity: Airborne particles can introduce flaws, obstruct nozzles, and deteriorate materials.

7. Commitment to Evolution and Constructive Criticism

Adapting and refining is the path to perfection.

Stay Abreast: Embrace ongoing developments in techniques and technology. This commitment ensures alignment with industry best practices.

Iterative Feedback Mechanism: Construct a cyclical process to record, review, and learn from every print experience, be it positive or challenging.

In summation, delivering consistently exceptional print quality is an orchestration of insightful material choices, equipment familiarity, meticulous parameter settings, and fostering the right environment. By mastering these dimensions and cultivating an ethos of perpetual learning, each print will resonate with unmatched precision and artistry.

## Diagnosing and resolving advanced print issues

Within the multifaceted world of modern printing, which spans both 2D and 3D technologies, proficiency is more than just mastering machine operations. It's about deeply understanding how to spot and address issues. While some printing problems can be straightforward, advanced ones require a more detailed and comprehensive approach. This guide delves into various complex print challenges and offers guidance on pinpointing and resolving them.

1. Unintended Shape Alterations in 3D Prints

Identified by: Shifts between layers, inconsistent adhesion, or deviations from intended geometry.

Possible Causes:

Machine Discrepancies: Misaligned beds, slackened belts, or calibration errors can contribute.

Filament Irregularities: Exposure to temperature swings or humidity can lead to inconsistent layers.

Slicer Miscalculations: Erroneous configurations can affect the final print.

Solutions:

Schedule regular equipment checks to ensure belt tension and smooth rail movements.

Use dry, quality filaments and stabilize printer temperature settings.

Reconfigure slicing software according to the specific printer and filament.

2. Color Deviations in 2D Prints

Identified by: Lines across the print, uneven coloration, or hues not matching the original.

Possible Causes:

Ink Issues: Blocked print nozzles or outdated ink can diminish print quality.

Calibration Disparities: Inconsistent color setups across devices can introduce color shifts.

Paper-Ink Incompatibility: Using inks incompatible with certain paper types.

Solutions:

Periodically clean print nozzles and use inks within their recommended shelf life.

Ensure devices share consistent color settings.

Match inks with compatible paper types.

3. Fragility in 3D Printed Structures

Identified by: Breakable parts, weak layer bonds, or unexpected internal gaps.

Possible Causes:

Extrusion Issues: Blockages or calibration errors can affect material deposition.

Temperature Instability: Inappropriate temperatures can negatively impact layer bonds.

Slicer Settings: Incorrect configurations can lead to voids.

Solutions:

Regularly inspect and adjust the extruder for smooth filament output.

Adjust temperature settings based on the filament type and ambient conditions.

Modify slicer configurations to optimize internal part structures.

4. Imperfect Surface Quality in Prints

Identified by: Unwanted surface textures or irregularities in both 2D and 3D prints.

Possible Causes:

Material Contaminants: Impurities in filaments or inks can degrade surface quality.

Travel Settings: Inefficient nozzle movements can result in 3D print anomalies.

Surface Inconsistencies: Debris or irregularities can impact the end product.

Solutions:

Always source and store materials to prevent contamination.

Adjust slicer settings to optimize nozzle paths in 3D prints.

Keep the printing surface or paper tray clean and level.

5. Attachment Challenges in 3D Printing

Identified by: Edges lifting off, print deformation, or failure to stick to the bed.

Possible Causes:

Surface Preparation: Inappropriate bed surfaces or adhesion methods.

Cooling Dynamics: Inconsistent cooling rates or wrong bed temperatures.

Solutions:

Experiment with various bed treatments and adhesion aids.

Monitor and adjust cooling rates and bed temperatures for optimal adhesion.

In wrapping up, while complex print issues can be challenging, they are not insurmountable. By adopting a methodical troubleshooting approach, paired with routine maintenance and continuous learning, it becomes feasible to consistently produce high-quality prints.

## Upgrades and modifications for print quality

Attaining superior print results in both 2D and 3D realms demands a harmonious blend of sophisticated equipment, updated software, and precision-based methodologies. As the expectations for unparalleled quality surge, enthusiasts and industry professionals continually seek methods to amplify their machines and techniques. This exploration presents a detailed look at essential enhancements and adaptations crucial for augmenting print quality.

1. Refining the 3D Printer's Hardware Components

Elevating Extruder Mechanics: The core of material layering, an extruder can benefit from precision-focused refinements. Transitioning to a dual-gear mechanism can assure consistent filament movement, curbing common challenges like filament jams or inconsistent extrusion.

Advanced Bed Heating Mechanisms: A key to combating warping and ensuring adherence is a high-end heated bed. Materials like PEI coatings or borosilicate surfaces can promise not only excellent adhesion but also effortless post-print handling.

Precision-Driven Rails and Bearings: Adopting top-notch linear rails and bearings promises fluidic and pinpoint movements, resulting in intricately detailed prints with negligible layer variances.

2. Software-Driven Enhancements for Pristine Outputs

Slicing Software Mastery: The bridge between design blueprints and tangible prints is the slicing software. Venturing into its intricate settings—like dynamic layering, innovative infill patterns, or top-layer refinement—can drastically optimize print outcomes.

Timely Firmware Revisions: Regular firmware updates synchronize the printer with emerging software capabilities, enhancing both the machine's responsiveness and the eventual print quality.

Refinement Tools for Digital Models: Leveraging applications like Meshmixer or PrusaSlicer aids in perfecting 3D designs, rectifying digital glitches, and gearing them for optimal printing.

3. Choosing Premium Materials for Distinguished Outputs

Selecting Elite Filaments: A conscious choice of top-tier filaments guarantees predictable and exceptional print results. Uniform diameter, purity, and appropriate storage are vital aspects of this selection.

Adopting Renowned 2D Inks: For the 2D spectrum, gravitating towards pigment-focused or long-lasting inks can accentuate color depth, and durability, and sidestep issues like untimely fading.

4. Maintaining Environmental Equilibrium for Uniform Results

Shielded Printing Zones: To cater to materials sensitive to environmental fluctuations, a shielded printing chamber can help maintain consistent internal conditions, ensuring the material behaves uniformly.

Monitoring Vital Environmental Parameters: Incorporating temperature and humidity tracking devices can alert you to any abrupt changes that might jeopardize print quality. This becomes especially pivotal for certain materials in both the 3D and 2D printing realms.

5. Mechanical Alterations for Heightened Accuracy

Optimizing Belt Dynamics: Introducing adjustable components to maintain belt tension ensures movement precision and diminishes issues like misalignment or layer shifting.

Stabilizing the Z-Axis: For prints demanding height or printers with extended z-axis capabilities, adding stabilizers can eradicate undesirable movements, promising even layering.

Diverse Nozzle Options: Depending on the project specifications or the chosen material, swapping to an alternate nozzle can significantly impact the granularity and durability of the print.

6. Calibrating with Finesse

Automated Bed-Leveling Mechanisms: While manual leveling poses challenges, automated kits simplify the process, ensuring a well-aligned base at the outset of every print.

Proactive Filament Detection: Integrating sensors that can preemptively flag filament shortages or inconsistencies can save a print from unexpected failures.

In wrapping up, the path to unmatched print quality is one of continuous learning and adaptation. A synergy of equipment enhancements, software updates, and material choices paves the way for unmatched results. As printing technologies evolve, staying updated with the newest developments remains crucial for those aiming for the zenith of quality.

# Chapter Twelve

## Software and Firmware Mastery

### Advanced slicing settings and features

Slicing software stands at the crossroads between a digital design and its physical counterpart in the world of 3D printing. With the relentless evolution of 3D printing techniques, slicing platforms have stepped up, introducing sophisticated settings and features that help ensure quality, streamline the process, and economize on materials. By harnessing these advanced configurations, designers and technicians can fine-tune print approaches, amplifying the excellence of the resulting print.

1. Dynamic Layering:

Conventional slicing divides a model into even layers. Advanced slicing introduces dynamic layering, altering layer thickness in tandem with the model's structure. Complex sections benefit from finer layers, while straightforward areas can be handled with broader ones, striking a balance between speed and detail.

2. Diverse Infill Shapes:

Moving beyond the typical infill configurations such as rectangular, triadic, or hexagonal, advanced slicing platforms offer specialized patterns like gyroid or three-dimensional subdivisions. These patterns enhance structural robustness, minimize material consumption, and can even adjust properties like resonance or fluid dynamics of a part.

3. Surface Refinement:

Surface refinement techniques slightly remelt the topmost layer, offering a polished finish, effectively diminishing the appearance of individual layers and negating the need for additional refinement post-print.

4. Perimeter and Surface Line Adjustments:

Fine-tuning the breadth of lines can fortify external perimeters or bolster the cohesion of the topmost and base layers. Sophisticated settings might even facilitate multiple perimeter widths in a single output.

5. Concentric Infill:

This strategy refines the inner density of a model. It bolsters density nearer the exterior while lightening it centrally, ensuring a sturdy shell with efficient core material usage.

6. Precision in Hole Expansion:

Achieving pinpoint accuracy for hole dimensions post-print can be tricky due to material contractions. Adjusting horizontal expansions can fine-tune these measurements.

7. Bridging Configurations:

Bridging refers to executing flat segments between elevated sections sans supports. Tweaking parameters such as material flow, pace, and thermal management guarantees stable and droop-resistant bridges.

8. Environment Shields:

For substrates that are susceptible to external conditions, an environment shield erects a minimal barrier around the piece, fostering a consistent internal climate and safeguarding from stray airflows, especially beneficial for materials like ABS.

9. Seam Management:

Seams, or the points where layers commence and culminate, can be conspicuous. Advanced tools empower users to dictate the seam's position or scatter it, making it less prominent.

10. Navigation and Avoidance Strategies:

Sophisticated slicers provide methods like "Evade Printed Sections" or "Combing" to steer clear of traversing printed portions during travel, eliminating potential surface blemishes.

11. Adaptive Speed and Motion Handling:

Adaptive configurations let the printing apparatus adjust speed and motion metrics depending on the piece's form, enabling fluid arcs and precise angles, thus optimizing both momentum and print finesse.

12. Support Barrier Layers:

While supports underpin overhanging parts, extracting them can sometimes affect the surface. Barrier layer configurations produce a dense but detachable barrier between the model and its support, yielding a pristine finish after extraction.

13. Handling Multiple Materials and Extruders:

Advanced software can control multi-extruder apparatuses, facilitating detailed multi-material outputs. These settings harmonize material transitions, waste towers, and protective barriers for clarity and efficiency.

14. Tailored Scripting:

For those adept with G-code, slicing platforms offer tailored post-processing scripting, enabling automation, bespoke movements, and custom device behaviors beyond preset parameters.

15. Component-Specific Modifiers:

Instead of a one-size-fits-all approach, component-specific modifiers let users tailor settings for distinct sections of a model, say, to amplify thickness or density where there might be more stress.

In essence, the plethora of configurations at the disposal of users in contemporary slicing platforms can vastly elevate the 3D printing journey. Each feature, aimed at resolving specific challenges, can, when mastered, transform the quality and efficiency of the printing experience.

## Custom firmware and printer modifications

Navigating the realm of 3D printing often leads to a quest for enhanced performance and precise customization. Hardware and material advancements certainly garner attention, but it's the often-understated firmware that plays a pivotal role behind the scenes. For those intent on optimizing their printing experiences, delving into the intricacies of custom firmware and specialized printer alterations can open new avenues of capabilities and fine-tuning.

1. Understanding Firmware's Role:

Think of firmware as the bridge between software commands and hardware executions. This layer dictates a printer's every move, from heating dynamics to extrusion behavior. Although default firmware serves most users adequately, custom iterations offer a deeper realm of control and adaptability.

2. What Custom Firmware Brings to the Table:

Adaptable Printing Profiles: From energy conservation modes to hushed operations, custom firmware can cater to unique printing needs.

Superior Accuracy: Improved motion algorithms can yield refined print textures, optimal layer bonding, and minimal printing flaws.

Robust Security Protocols: Embed custom security measures to prevent unwarranted machine access or mitigate operational errors.

Compatibility with Additional Hardware: Adjust firmware to seamlessly incorporate and optimize new hardware enhancements.

3. Top Picks in Custom Firmware:

Marlin: A darling of the open-source community, Marlin is feature-rich, frequently updated, and versatile enough for numerous printer setups.

Repetier: Appreciated for its intuitive nature, Repetier offers options like multi-extruder integration, tailored printer geometry, and built-in G-code analytics.

Klipper: Gaining traction for its innovative approach, Klipper delegates motion computations to an external computer, promising faster operations and sharper accuracy.

4. Printer Modification Explained:

Beyond firmware, direct hardware augmentations can profoundly shape a printer's functionalities, from foundational tweaks to advanced reconfigurations.

Upgraded Hotend and Extruder Systems: Premium hotends maintain even heat distributions, and superior extruders ensure filament consistency, broadening material compatibility.

Advanced Print Beds: Transitioning to specialized beds like magnetic, flex-oriented, or textured versions can simplify post-print processes and enhance material adherence.

Revamped Cooling Mechanisms: Redesigned fans or air channels guarantee optimized cooling, crucial for intricate printing sections.

Noise-Reducing Stepper Drivers: For those prioritizing discretion, these drivers dramatically cut down operational sounds.

Automated Bed Calibration Tools: Bypass manual calibrations with automated systems that guarantee uniform first-layer prints.

Responsive Filament Monitors: These units can detect filament anomalies, pausing operations to allow timely interventions.

5. Navigating Potential Hurdles:

Embracing custom firmware and hardware adjustments warrants careful consideration. Mismatched firmware could compromise equipment or diminish print outcomes. Similarly, certain hardware tweaks might affect equipment warranties or necessitate recalibrations.

Always Maintain Backups: Prior to initiating modifications, ensure you've safeguarded original firmware and configurations, offering a safety net against unexpected outcomes.

Maintain Comprehensive Records: Detailed logs of changes, particularly with firmware adjustments, can be invaluable for future references or problem-solving.

Engage with Enthusiast Communities: Gleaning insights from dedicated forums or groups can shed light on best practices and potential pitfalls.

Adopt a Step-by-Step Approach: Undertake and evaluate changes gradually, enabling precise identification of any issues that might arise.

In Closing:

The journey through custom firmware and printer modifications can transform a standard 3D printing environment into a tailored powerhouse of efficiency and innovation. Yet, this potential demands a meticulous approach, where careful planning, consistent backups, and comprehensive documentation are paramount.

## Scripting and automation for 3D printing

The merger of 3D printing with the worlds of scripting and automation heralds a significant shift in the manufacturing sector. By synergizing automated command sequences with intricate fabrication mechanics, the horizons of precision, consistency, and bespoke outputs have expanded exponentially.

Understanding the Role of Scripting:

At its core, scripting in 3D printing involves a predefined set of instructions aimed at accomplishing specific functions. Such functions could span from initiating printer calibrations and overseeing print sequences, to executing post-print activities like power-down procedures or sending completion alerts. The allure of scripting is twofold: it reduces hands-on interventions and guarantees that operations remain at a peak standard, irrespective of the task's intricacy.

Predominant Scripting Platforms in 3D Printing:

Python: Given its adaptability and comprehensive support libraries, Python has become a popular choice for scripting in 3D printing. Utilities like PyCAM and the plugins in Cura provide a conducive space for Python-based scripts targeting various operations.

G-code: Renowned as the default language for many 3D printers, G-code provides direct access to customize the print journey. This might involve tweaking layer profiles mid-operation or altering thermal dynamics on the go.

The World of 3D Printing Automation:

Beyond scripting, automation encapsulates the idea of system-wide cohesion, aiming to execute tasks with minimal human input. Automation in the 3D print sector can take several shapes:

Sequential Printing: For repetitive designs, automation tools can rejuvenate the printer after each cycle, facilitating uninterrupted production.

Self-Leveling Platforms: Instead of manual bed adjustments, modern tools combined with scripts autonomously guarantee a consistently level platform.

Filament Monitoring and Swap-Outs: Today's printers, fortified with sensory mechanisms, can halt operations during filament depletion, triggering automated alerts or replacements.

Remote Oversight and Manipulation: The surge in connected devices allows printers to be supervised and adjusted from afar. Automated scripts can mitigate unforeseen disturbances, relay operation snapshots, or modify settings based on live feedback.

Scripting Meets Automation:

While individually potent, the amalgamation of scripting with automation promises unmatched operational efficiency. Envision a situation where filament obstruction is sensed. A holistic system could freeze the operation, manage the filament, alert the operator, and after addressing the issue, restart the print—each step orchestrated by a combination of scripts and automation.

Potential Pitfalls and Precautions:

While the appeal of such integration is evident, certain challenges must be navigated:

Inherent Complexity: Scripting necessitates a grasp of coding principles. Although numerous ready-to-use scripts are accessible, personalizing them might need deeper expertise.

Equipment Constraints: Advanced scripts or automation features might not be congruent with all printer models, especially antiquated ones. Verifying hardware support is crucial before delving into complex scripting.

Dependency Dangers: Automation does enhance workflow, but excessive dependence might lead to neglecting mechanical anomalies or sporadic quality verification.

Wrapping Up:

Integrating scripting and automation into the 3D printing realm signifies the industry's metamorphosis. Be it individual enthusiasts or large-scale industrial units, the aspiration for accuracy, effectiveness, and consistency is paramount. The confluence of digital strategies with mechanical operations signals the dawn of a transformative era in additive manufacturing.

# Chapter Thirteen

## 3D Printing with Robotics and Automation

### Integrating 3D printing with robotic arms and systems

In the fast-paced world of manufacturing, 3D printing stands out as a groundbreaking advancement. But, the potential of this technology doesn't end there. Pairing 3D printing with the capabilities of robotic arms and systems takes this innovation to even loftier heights, offering unparalleled precision, flexibility, and scale in additive manufacturing endeavors.

Why Merge the Two Technologies?:

3D printing shines in its ability to bring digital designs to life in the physical realm. When you combine this with the dexterity and range of robotic systems, you're looking at an even more dynamic and versatile manufacturing method. The multifaceted nature of robotic arms means they can be programmed to handle print nozzles in more complex ways, leading to wider printing patterns, diverse spatial orientations, and the potential for multiple simultaneous prints.

The Benefits of this Integration:

The union of 3D printing with robotic technology introduces several key enhancements:

Directional Flexibility: Traditional 3D printing follows a set path, layering from the bottom up. Robotic arms break this mold, offering varied printing directions that can produce intricate patterns and shapes without additional support.

Versatility in Functions: Robotic systems don't just stop at printing. Once the printing phase is complete, these robots can shift gears to handle other tasks, from refining the product to assembly, and optimizing the production process.

Greater Scale in Production: Standard 3D printers are bound by size constraints, but robotic systems can handle much larger projects. This is especially useful in sectors like construction or automotive manufacturing.

Diverse Material Use: The precision of robotic systems allows for switching between different materials during a print. This paves the way for items made of multiple materials or varied material consistency.

Challenges and Considerations:

However, integrating these two technologies isn't without its complexities:

Fine-tuning the Process: Achieving a balance between the robot's movement and the precision of 3D printing requires keen attention to calibration. This includes ensuring accurate movement, adjusting material flow, and accounting for heat factors.

Software Limitations: 3D models need to be converted into commands for the robotic system. This often calls for dedicated software solutions, which are still in their developmental stages.

Initial Costs: Investing in advanced robotic systems isn't cheap. When combined with 3D printing tech, it can present a significant financial hurdle, especially for smaller operations.

Exploring Applications:

The combined tech offers a broad spectrum of applications:

Building and Infrastructure: Imagine robotic arms aiding in the on-site production of building components, or even entire houses using durable materials.

Tailored Production: In areas like custom car parts or prosthetics, robotic precision can create items perfectly suited to individual needs.

Artistic Endeavors: Artists can utilize the enhanced printing capabilities to craft intricate pieces, pushing creative boundaries.

Peering Ahead:

The combination of 3D printing and robotic systems paints a promising picture for manufacturing's future. Envision an environment where printers aren't stationary but move around a workshop, collaborating with other systems, and even undertaking self-maintenance.

To sum it up, while the journey of combining 3D printing with robotic systems has its challenges, its potential is vast. Offering greater efficiency, scalability, and versatility, this merger is poised to redefine manufacturing in the coming years.

## Automated assembly and post-processing

The landscape of modern manufacturing has seen a significant evolution, with automation emerging as a cornerstone of enhanced efficiency and precision. Notably, these advances aren't confined to the initial production stages but have seamlessly permeated into the assembly and post-processing realms. By harnessing automated assembly and post-processing techniques, we can ensure unparalleled quality, mitigate human errors, and amplify productivity.

The Case for Automated Assembly:

At its core, automated assembly leverages technologically advanced devices, such as robotic arms or sophisticated actuators, replacing tasks traditionally done by human hands. This pivot towards automated solutions is underpinned by several compelling factors:

Expediency and Productivity: Mechanical systems frequently surpass human assembly capabilities, especially concerning repetitive operations.

Accuracy and Uniformity: Robotic mechanisms uphold a degree of exactitude that is challenging for humans, guaranteeing product consistency.

Enhanced Safety Protocols: Tasks that involve handling potentially hazardous materials or repetitive actions can be injury-prone. Automation dramatically curtails these risks.

Key Facets of Automated Assembly:

Innovative techniques and components facilitate the streamlined operation of automated assembly:

Advanced Sensing Technology: These tools, including vision and infrared systems, are pivotal for pinpointing the alignment and positioning of components, and adjusting robotic actions when needed.

End Effectors: These are the robotic system's functional extensions, be it grippers, specialized tools, or suction mechanisms.

Real-time Feedback Mechanisms: These systems continually oversee the assembly process, relaying instantaneous data to ensure tasks remain accurate and efficient.

Venturing into Automated Post-Processing:

Post-production processes, spanning from cleaning to inspection, gain much from automation, primarily due to:

Standardized Outcomes: In operations like painting or surfacing, automation guarantees a uniform finish.

Optimized Resource Utilization: Accurate control tools enhance material use, minimizing excesses.

Flexibility in Operations: Automated post-processing can adeptly adjust based on production requisites, sidestepping the hurdles of manual methods.

Pioneering Tools in Automated Post-Processing:

Automated Transport Systems: Mobile robotic units in expansive manufacturing settings facilitate seamless movement between workstations.

Computer-Driven Machining Tools: Predominantly used for detailed tasks such as carving or drilling, these machines offer unparalleled precision.

Automated Inspection Platforms: With the infusion of AI-powered algorithms, these systems can swiftly and accurately identify anomalies or defects.

Potential Hurdles:

While automation brings a plethora of advantages, there are inherent challenges:

Upfront Costs: Establishing these automated platforms requires substantial financial investments.

Routine Upkeep: Over time, maintenance becomes imperative, and any disruptions can impede the operational flow.

Requisite Skill Transitions: Employees must undergo training to operate, oversee, and troubleshoot these advanced systems.

Envisioning What Lies Ahead:

Future trajectories in AI, interconnected devices, and robotic capabilities suggest a vast potential for automated systems in both assembly and post-processing. Integration with concepts like the Internet of Things (IoT) predicts these platforms will become more collaborative, self-regulating, and dynamic.

To wrap up, as global industries chase after enhanced output, cost reductions, and impeccable product quality, automated assembly, and post-processing emerge as integral components. These methods, characterized by their precision and adaptability, are no longer merely future constructs—they define our current reality. By capitalizing on these technologies, businesses can remain at the forefront, primed to address the dynamic demands of today's global consumer base.

## Case studies of 3D printing in modern factories

Over recent years, 3D printing, also known as additive manufacturing, has shifted from being a specialized prototyping method to a cornerstone of today's production landscapes. Its capabilities for swift prototyping, individual customization, and just-in-time production have captured the attention of various sectors. This exploration delves into several case studies illustrating the pivotal role of 3D printing in today's manufacturing scenarios.

1. Revolutionizing the Aerospace Domain

The aerospace sector has been at the forefront of embracing 3D printing, driven by strict regulations and the perpetual quest for weight-efficient, robust components.

Highlight: GE Aviation

Through additive manufacturing, GE Aviation fashioned a fuel nozzle for their LEAP engines. While traditionally this piece was made of several parts, 3D printing enabled a unified, stronger structure. This innovation resulted in a weight cut by a quarter and significantly enhanced component longevity.

2. Driving Change in the Automotive Realm

The world of automobiles has leveraged 3D printing to streamline prototype creation and enable efficient component customization.

Highlight: BMW Group

BMW has integrated additive manufacturing for vehicle development and mainstream production. In a notable initiative, their MINI Yours Customized program empowers customers to design specific vehicle elements. These designs, ranging from decorative elements to functional parts, are then brought to life via 3D printing.

3. Healthcare: Crafting Custom Solutions

The medical arena's adoption of 3D printing, especially for personalized solutions like prosthetics or dental aids, has been transformative.

Highlight: Align Technology

Align Technology, the creators of Invisalign braces, use 3D printing for producing individualized dental aligners. By digitally mapping patients' dentition, they produce a myriad of unique molds each day, ensuring both comfort and effective treatment.

4. Footwear Industry: Merging Aesthetics with Comfort

Footwear manufacturers are tapping into 3D printing to craft shoes that balance design with personalized comfort.

Highlight: Adidas

Joining hands with the 3D printing enterprise Carbon, Adidas launched the Futurecraft 4D footwear line. The shoes' soles, crafted using advanced printing techniques, offer both durability and tailored ergonomics, catering to athletes' specific needs or general customer preferences.

5. Building the Future in Construction

While the construction industry has been traditionally resistant to rapid tech adoption, 3D printing is making its mark, particularly for infrastructure and housing projects.

Highlight: Winsun

Winsun, a firm from China, garnered attention by constructing ten houses within a day using 3D printing. By blending cement with recycled construction debris, they printed building elements, which were later assembled on-site, signifying savings in time, resources, and finances.

6. Culinary Innovations with 3D Printing

Surprisingly, the culinary sector is also experimenting with 3D printing, especially for creating detailed designs.

Highlight: Choc Edge

Choc Edge has introduced a 3D printer specifically for chocolate, granting chefs the ability to craft detailed and unique designs. This venture into food printing could pave the way for personalized nutrition, tailoring meals to individual dietary requirements.

Wrapping Up:

These instances merely scratch the surface of 3D printing's profound influence on contemporary manufacturing processes. The ramifications extend beyond production to aspects like supply chain evolution, workforce upskilling, and product positioning. With continuous technological advancements, 3D printing is poised to further redefine manufacturing, marking an era of precision, adaptability, and unmatched efficiency.

# Chapter Fourteen

## Sustainable and Green 3D Printing

### The environmental impact of advanced materials

In today's industrial landscape, the adoption of advanced materials like composites, nanomaterials, and biomaterials is increasingly prevalent. Their unmatched capabilities cater to sectors ranging from aviation to health. But as we embrace these innovations, it's vital to understand their ecological implications. Here's a nuanced look at the environmental considerations surrounding advanced materials.

1. Origin and Production

Every material's journey starts with its creation, often demanding resource extraction. Take carbon fiber composites, which lean on polyacrylonitrile (PAN) from petrochemical origins. Extracting and refining these resources come with well-charted environmental challenges – think habitat disturbance, pollution, and the emission of greenhouse gases.

Moreover, crafting some of these advanced materials requires high-energy processes. High-temperature ovens for ceramics and composites, for instance, lead to significant energy draws. When coupled with the use of non-renewable energy sources, this heightens the carbon footprint of these materials.

2. The Nanomaterial Conundrum

While nanomaterials, with their impressive features, herald new technological eras, their interaction with the environment is a budding area of study. Some evidence indicates the potential toxicity of specific nanomaterials to marine organisms and their potential accumulation in ecosystems. Their minuscule size allows them to infiltrate cells, posing risks of cellular harm. Plus, their expansive reactive surface means they might engage in detrimental environmental reactions.

3. Decomposition and Environmental Persistence

A significant concern with many state-of-the-art materials is their resistance to biodegrading. While polymer composites boast longevity and resilience, they can linger in the environment long after their utility ends, adding to the mounting issue of enduring waste.

Conversely, some biomaterials, designed for health applications, emphasize biodegradability. Take polylactic acid (PLA), a material from renewable sources that can break down over time, thus limiting environmental damage. Yet, it's crucial to recognize that biodegradability isn't a universal trait of all biomaterials.

4. Recycling Dilemmas

Recycling, a cornerstone of sustainable waste management, faces hurdles with advanced materials. For instance, fiber-reinforced composites don't readily melt and reshape like conventional plastics. Efforts to retrieve the matrix or fibers are complicated and energy-draining.

However, the tide is changing. Research is underway to create recyclable composites and refine recycling practices. Industries heavily reliant on these materials are pouring resources into enhancing product recyclability post-usage.

5. The Silver Lining: Energy and Emissions

It's only fair to also shed light on the positive environmental facets of advanced materials. Their lightweight attributes, especially in sectors like aviation and transport, lead to notable energy savings and decreased emissions. The impressive strength-to-weight metrics of materials like carbon composites have lightened aircraft, optimizing fuel consumption.

Also, in the construction realm, cutting-edge insulation materials are elevating building energy efficiency, slashing the need for heating and cooling, and cutting down emissions.

6. Steps Forward

Awareness of the environmental challenges tethered to advanced materials is spawning various proactive measures:

Crafting materials through environmentally sensitive methods.

Sourcing sustainable raw materials.

Perfecting recycling techniques for end-of-life materials.

Amplifying research on the eco-impact of materials, particularly at the nano level.

Wrapping Up:

While advanced materials have propelled sectors into new frontiers of innovation and functionality, their environmental aspects call for introspection. Incorporating green practices at every stage of a material's life, from creation to disposal, is imperative. As technological advancements surge, marrying them with a commitment to ecological stewardship will chart the course ahead.

## Techniques for waste reduction and recycling

As global waste continues to mount, largely propelled by increased consumer habits, the call for sustainable management, diminished waste production, and effective recycling grows louder. Even though achieving a world without waste might sound idealistic, cutting-edge techniques are inching us ever closer to this ambition. Let's delve into a comprehensive look at contemporary methods of waste reduction and recycling.

1. Prevention at the Source

Also termed waste deterrence, this strategy zeroes in on curbing the genesis of waste:

Conceptualization: Products are re-envisioned to use fewer resources while ensuring they last longer.

Manufacturing Adjustments: Processes are fine-tuned to limit raw material input and slash waste production.

Educating the Buyer: End-users are equipped with insights on eco-friendly consumption, encouraging choices that favor enduring products.

2. Modernized Sorting Mechanisms

For recycling to thrive, waste segregation is vital. Today's advanced systems utilize AI, specialized sensors, and robotic interventions to meticulously separate recyclable materials, thereby bolstering recycling effectiveness.

3. Evolved Mechanical Recycling

Traditional mechanical recycling processes like melting or reconstituting are undergoing evolution:

Polymeric Mixes: Merging assorted plastic wastes to enhance their combined attributes, making recycled materials more versatile.

Controlled Breakdown: Techniques are in place to minimize polymer wear during recycling, maintaining the integrity of the recycled product over several cycles.

4. Molecular Recycling

By breaking polymers into foundational molecules, they can be remodeled into fresh-quality polymers. This is particularly adept at handling mixed or tainted plastic waste:

Pyrolytic Techniques: In oxygen-free conditions, waste plastics are thermally decomposed into synthetic crudes, and subsequently reshaped into plastics or other derivatives.

Gasification Methods: Using thermal processes, waste is transformed into synthetic gases that are further employed to synthesize assorted chemicals and energy products.

5. Managing Organic Refuse

Domestic and industrial organic refuse can be mined for resourceful outputs:

Natural Decomposition: Speeding up natural decay to convert organic waste into agriculturally valuable compost.

Digestion without Oxygen: Microbes process organic refuse in sealed conditions, releasing methane-rich biogas usable for power generation.

6. Modern E-waste Solutions

Given the digital age, e-waste solutions are paramount:

Water-Based Metal Retrieval: Utilizing liquid chemical solutions to reclaim valuable metals like platinum, copper, and rare earth elements from discarded electronics.

Microbial Metal Recovery: Harnessing specific microbes that can extract metals from e-waste, offering an eco-friendly alternative to chemical methods.

7. Manufacturer-Led Waste Initiatives

Here, producers take charge of the end-of-life phase of their merchandise. By ensuring retrieval, recycling, or apt disposal, they play a key role in diverting waste from landfills.

8. The Circular Economy Model

Moving away from the traditional "consume and discard" model, the circular approach focuses on maximizing resource utility:

Assembly & Disassembly Designing: Fabricating products that can be effortlessly dismantled to facilitate efficient component recycling or reuse.

Service-Centric Business Models: Companies pivot to offer services like leasing or refurbishing, which prolong product usage and reduce waste.

9. Active Consumer Involvement

Raising public awareness about waste management, the merits of conservative consumption, and recycling's significance is crucial. Grassroots movements, training sessions, and digital outreach are fundamental in molding eco-conscious habits.

10. Regulatory Boost

Robust waste management norms, curbs on disposable plastics, setting ambitious recycling goals, and incentivizing green business tactics can act as catalysts for profound change.

Wrapping Up:

Tackling the waste challenge demands a blend of pioneering methods, reshaped societal values, and adaptable business structures. With the confluence of these innovative techniques and a responsible populace, we can chart a course toward a future where every resource is treasured and waste is a rarity.

## Sustainable energy sources for 3D printing

In today's industrial landscape, the momentum behind sustainable practices intersects with the expansive growth of 3D printing. Though 3D printing has revolutionized manufacturing by offering design agility, customized production, and swift prototyping, the energy demands of these systems have become a focal point of discussion. This exploration delves into the integration of eco-friendly energy solutions with 3D printing mechanisms.

1. Tapping Solar Power

The sun's energy, captured using photovoltaic systems, has become an increasingly viable solution with a consistently reducing cost trajectory. By situating solar infrastructure adjacent to or within manufacturing facilities, the energy demands of 3D printers can be aptly addressed. To counter the variable nature of solar energy, having a competent energy storage mechanism ensures continuous printing operations even during sub-optimal sunlight periods.

2. Harnessing Wind Power

Advancements in wind turbine design have led to remarkable energy yields. Choosing locations with steady wind currents to set up 3D printing infrastructure can effectively utilize this renewable energy. Given the occasional variability in wind intensity, integrating powerful energy storage systems remains a prudent strategy.

3. Employing Bioenergy and Biogas

Energy derived from organic waste, including agricultural leftovers, forestry remnants, or dedicated energy crops, can be converted into usable electricity or thermal energy. Methane-rich biogas, a product of anaerobic decomposition, serves as a potential energy reservoir for 3D printing operations. Adopting bioenergy aligns 3D printing with a green and sustainable energy cycle.

4. Micro-Hydropower Systems

While dams dominate hydropower narratives, micro-hydropower solutions have carved a niche in localized energy generation. Capitalizing on water flow, such systems can offer uninterrupted energy for 3D printers, especially if based near perennial water streams.

5. Delving into Geothermal Sources

By leveraging the Earth's core heat, geothermal systems present an alternative for locations with natural geothermal propensities. Geographical constraints might limit its universal adoption, yet areas with inherent geothermal activity can ideally serve as 3D printing clusters powered by this renewable source.

6. Fine-tuning 3D Printing Energy Consumption

Parallel to sourcing renewable energy, the focus on energy-efficient 3D printing is of the essence:

Adaptive Energy Consumption: Contemporary 3D printers often incorporate adaptive energy management features. Such systems adjust energy intake depending on the printing task, optimizing overall usage.

Material Breakthroughs: Evolving research has introduced materials that mandate lower energy during the printing process, reducing the overall energy imprint.

Streamlined Printing Pathways: Computational techniques can devise energy-optimal printing trajectories, minimizing redundant motions and conserving energy.

7. Renewable Grids and Certifications

For setups unable to directly adopt renewable energy, plugging into an eco-friendly grid remains an option. Several energy providers offer renewable energy certificates, essentially certifying that the acquired energy originates from green sources, albeit indirectly.

8. A Panoramic View of Sustainability

Beyond energy, a rounded approach to sustainability in 3D printing would consider eco-friendly material procurement, waste material recycling, and designing products with an emphasis on durability and minimalism.

In Summation:

The amalgamation of sustainable energy sources and 3D printing represents more than a trend—it's a mandate in a planet confronting ecological challenges. Embracing green energy in 3D printing amplifies the sector's contribution to global environmental preservation. As the fields of renewable energy and 3D printing evolve, their partnership heralds an era of innovation, efficiency, and overarching sustainability.

# Conclusion

## Reflection on the advanced journey of 3D printing

From its inception in the 1980s, 3D printing's transformative path has been a tale of continuous innovation. What began as a tool to bring digital designs into the tactile realm has evolved into a multifaceted manufacturing powerhouse.

Initially perceived as a mechanism for rapid prototyping, the first 3D printers served primarily as devices for design visualization. Their primary aim was to hasten the product development phase rather than to provide comprehensive manufacturing solutions. Groundbreaking processes such as Stereolithography (SLA) were revolutionary, giving birth to the concept of producing tangible objects layer by layer from digital designs.

Despite its groundbreaking start, the early days were filled with hurdles. Costs were high, and materials, mainly resins, didn't possess the resilience that today's materials boast. But visionaries saw the expansive potential in 3D printing and believed that it could transcend mere prototyping.

The next decade saw an explosion of methodologies and materials in the 3D printing arena. The introduction of Fused Deposition Modeling (FDM) and Selective Laser Sintering (SLS) broadened its scope. The addition of metals, ceramics, and an array of polymers to the printing repertoire enabled not only prototypes but also operational components.

The synergy of evolving digital design tools and enhanced 3D printing technology marked a pivotal moment around the 2000s. It was a period of unhindered creative expression. Traditional manufacturing processes, with their inherent limitations, were challenged by 3D printing's additive approach, enabling the creation of complex designs and unparalleled customizations.

The rise of desktop 3D printers during this era signified a profound shift. No longer was 3D printing confined to industrial realms; it was now accessible to businesses, educators, and enthusiasts alike. This broad accessibility birthed a plethora of open-source initiatives and burgeoning communities, fueling further advancements.

Yet, it was the industrial sector's acknowledgment of 3D printing as a game-changing tool that truly marked its coming of age. Industries like aerospace, with their quest for lightweight and resilient components, became significant advocates. Medical sectors saw unmatched potential in bespoke solutions like tailor-made implants. Even artistic sectors, including fashion and architecture, began exploring 3D printing's potential, leading to groundbreaking creations.

As we recount this journey, it's essential to recognize the associated challenges. The ease of digital design replication raised questions about intellectual property. Environmental and energy concerns also came to the forefront. However, with the spirit that has always driven this field, these challenges are being met with innovative solutions, such as eco-friendly materials and energy-efficient methods.

To conclude, 3D printing's progression paints a vivid picture of human determination, adaptability, and visionary thinking. Its integration across sectors, from automobiles to gastronomy, stands as a testament to its transformative capacity. The journey thus far not only celebrates past milestones but also illuminates a promising path forward.

## Preparing for the future: what's next in the 3D printing world

As we propel into a future where 3D printing is a standard, one must ponder the transformative potential this technology holds. As 3D printing graduates from being a specialty process to an integral component of industries, it's intriguing to imagine the subsequent frontier. Let's delve into the unfolding chapters of this technology.

1. Broadening the Material Horizons

The forthcoming era will usher in an even wider array of printing materials. Beyond the staple polymers and metals, expect the dawn of multi-material prints. These could produce single objects with diverse attributes. Moreover, the evolution of smart materials will allow objects to alter properties when influenced by external factors like temperature or light.

2. The Medical Marvels of Bioprinting

The realm of bioprinting, or printing using biological materials, is brimming with promise. The vision of printing fully functional human organs tailored to individual genetic makeups could revolutionize organ transplants. Furthermore, the bespoke production of medical aids like prosthetics, dental solutions, and tailored medications is fast becoming a tangible reality.

3. Efficiency and Precision Combined

With technology continuously pushing boundaries, anticipate future 3D printers that merge the nimbleness of 3D printing with the efficiency of traditional methods. Concurrently, technologies such as two-photon lithography are a testament that high-speed production can still maintain incredible detail.

4. A Marriage of Software and Intelligence

As software innovations keep pace, 3D printing tools will increasingly embed AI capabilities. Such tools could proactively identify potential design pitfalls, adapt printing parameters dynamically, and even evolve designs based on user trends.

5. Symbiosis with Broader Technological Ecosystems

The integration of 3D printing within comprehensive manufacturing systems will redefine production efficiencies. Imagine scenarios where robots seamlessly manage post-printing processes, thus expanding the horizons of achievable outcomes.

6. The Dawn of Localized Production Units

Central to 3D printing's ethos is the principle of decentralized manufacturing. Instead of vast factories, the vision is community-centered print stations catering to localized demands, translating to minimized inventory and transportation overheads. This model presents both economic advantages and a potential reduction in environmental footprint.

7. A Commitment to the Planet

With an increasing global focus on eco-friendly practices, 3D printing stands at the forefront of responsible manufacturing. Inherent to its process is minimal waste generation. Moreover, the focus will likely shift towards harnessing recycled or earth-friendly materials.

8. New Avenues in Education

The impact of 3D printing will reverberate across educational domains. Its relevance will stretch beyond traditional tech disciplines, finding resonance in fields as diverse as medicine, arts, and even culinary arts.

9. Beyond Our Planet: Space Endeavors

One of the most captivating prospects is the potential role of 3D printing in space missions. As space agencies explore 3D printing in unique environments, we might soon witness the creation of extraterrestrial habitats using resources harvested from celestial bodies.

In essence, the impending phases of 3D printing promise a mosaic of innovations, influencing sectors far and wide. Like any evolutionary journey, obstacles will surface, be they ethical, ecological, or economic. Yet, challenges invariably foster creativity, ensuring that the future trajectory of 3D printing remains both vibrant and impactful.

## Encouraging continued exploration, innovation, and mastery

In today's rapidly changing technological environment, it's crucial to emphasize the value of perpetual exploration, innovation, and the drive for mastery. These components are the bedrock of advancement in every field, setting apart leaders from followers. Cultivating and upholding these qualities amidst continuous evolution requires careful and deliberate initiatives.

The Vitality of Ongoing Exploration

The commitment to unceasing exploration has been central to many of humanity's grand achievements. This innate curiosity has ushered in monumental discoveries and advancements.

Remaining at the forefront of any industry necessitates this restless drive to explore and push boundaries.

To nurture a culture of exploration:

Collaborative Ventures: Forming teams from varied backgrounds, initiating external partnerships, and fostering multi-disciplinary dialogues can be catalysts for fresh perspectives.

Education and Training: Introducing regular training sessions, seminars, and online courses can keep professionals curious and informed.

The Demand for Unwavering Innovation

Innovation isn't a mere choice; it's an imperative. It forms the linchpin between mere ideation and groundbreaking progress. In a world driven by competition, innovative minds lead the charge, setting the trajectory for the entire industry.

Steps to promote innovation encompass:

Fostering a Safe Environment: Encouraging an atmosphere where taking thought-out risks is seen positively can inspire groundbreaking ideas.

Acknowledgment and Incentives: Applauding and incentivizing innovative achievements can serve as a powerful motivator.

Commitment to Mastery

Mastery transcends mere expertise. It involves a dedication to continuously refining skills, aiming for unparalleled excellence, even when faced with adversity.

Approaches to champion mastery include:

Targeted Skill Refinement: Promote targeted and purpose-driven practices, focusing on honing specific skills.

Guidance Programs: Utilizing the wisdom of seasoned professionals can pave the way for emerging talent.

The Symbiotic Relationship among Exploration, Innovation, and Mastery

While each of these elements is powerful on its own, their combined force can lead to unparalleled outcomes. Exploration ignites curiosity, innovation converts this curiosity into tangible outputs, and mastery ensures the highest quality of these outputs.

Potential Hurdles in Upholding These Pillars

The merits of exploration, innovation, and mastery are evident. However, challenges like apprehensions about change or complacency with current methodologies can hinder progress. Addressing these obstacles requires visionary leadership and a culture that inherently values and champions these principles.

## Drawing Inspiration: The Trajectory of Tech Pioneers

Reflect on the evolution of industry titans such as Apple, Google, or SpaceX. Their success narratives highlight a robust combination of technological exploration, consistent innovation, and an unyielding quest for perfection. Their adherence to these principles has positioned them not just as industry leaders but as visionaries.

## In Summary

In a dynamic world, complacency is the enemy of progress. For those keen on making an impact, the path is clear: embrace continuous exploration, champion relentless innovation, and strive for absolute mastery. Such a commitment will foster individual and organizational evolution, ushering in a new era of growth and industry leadership.

Printed in Great Britain
by Amazon

45805758R00089